Synthesis Lectures on Ocean Systems Engineering

Series Editor

Nikolas Xiros, University of New Orleans, New Orleans, LA, USA

The series publishes short books on state-of-the-art research and applications in related and interdependent areas of design, construction, maintenance and operation of marine vessels and structures as well as ocean and oceanic engineering.

Alexander Arnfinn Olsen

Guidance on the Classification of Offshore Containers

 Springer

Alexander Arnfinn Olsen ⓘ
Southampton, UK

ISSN 2692-4420 ISSN 2692-4471 (electronic)
Synthesis Lectures on Ocean Systems Engineering
ISBN 978-3-031-74856-1 ISBN 978-3-031-74857-8 (eBook)
https://doi.org/10.1007/978-3-031-74857-8

This Springer imprint is published by the registered company Springer Nature Switzerland AG
The registered company address is: Gewerbestrasse 11, 6330 Cham, Switzerland

If disposing of this product, please recycle the paper.

Preface

The International Maritime Organisation (IMO) issued MSC/Circ.860 Guidelines for the approval of offshore containers handled in open seas. This circular is intended to assist the competent authorities in developing the requirements for approving the offshore containers. IMO requires that all intermodal containers conform to the requirements of the International Convention for Safe Containers (CSC). The requirements of the CSC convention may not be applicable to offshore containers primarily due to non-standard designs, exposure to the marine environment for extended periods as well as the lifting of offshore containers by pad eyes. EN 12079 has been published based on the MSC/Circ.860 and is currently used as an international industry standard to approve offshore containers. Containers built to Class standards for the certification of offshore containers will meet all the requirements of MSC/Circ.860, EN 12079:2006 and ISO 10855:2018. This Guide provides guidance for manufacturing facilities to build offshore containers. It also serves to assist Class Engineers and Surveyors in certifying offshore containers around the globe.

Southampton, UK   Alexander Arnfinn Olsen

The original version of the book has been revised. A correction to this book can be found at https://doi.org/10.1007/978-3-031-74857-8_16

Acknowledgements

It is with the deepest gratitude that I thank the team at Babcock Marine and Technology for sharing their knowledge and insights during the development of this guide. I would also like to extend my thanks to the editorial team at Springer for their assistance in bringing this guide together with particular thanks to Dr. Dieter Merkle and Prasanna Kumar Narayanasamy at Straive; and last but certainly not least, my deepest thanks go to my wife, who as always, has stood as steadfast as an anchor in the most violent of seas.

To you all, my sincerest thanks and gratitude.

Author's Note

Unless otherwise agreed in writing all services rendered and certificates issued in connection with offshore container and lifting set are governed by the terms and conditions established by each individual classification society and the offshore container/lifting set certification application ("Agreement") specification. By requesting offshore container/ lifting set certification, the client agrees to be bound by the terms and conditions set by Class. Whilst each individual classification society will have its own unique terms, conditions, and stipulations, it is possible to determine a standardised approach taken by the majority of classification societies.

Note: this chapter is intended as a summarisation of genericised Class terms and conditions and should not be treated as an authoritative instruction. Before applying for container/ lifting set certification, it is strongly recommended advice is sought from the appropriate Class engineering department.

Representations as to Certification

Certification is a representation by Class as to the structural fitness for a particular use or service in accordance with its specific Rules, Guides, and standards. The Class Rules and Guides are not meant as a substitute for the independent judgment of professional designers, naval architects, and marine engineers, nor are they intended as a substitute for the quality control procedures of shipbuilders, container manufacturers, steel makers, suppliers, manufacturers, and sellers of marine materials, machinery, or equipment. Indeed Class, being a technical society, can only act through Surveyors or others who are believed by it to be skilled and competent. Subsequently Class represents solely to the container/lifting set manufacturer, container/lifting set owner, or client of Class that when certifying it will use due diligence in the development of its Rules, Guides, and standards and in using normally applied testing standards, procedures and techniques as called for by the Rules, Guides, standards and other criteria of Class. Class further represents to the container/lifting set manufacturer, container/lifting set owner, or other client of Class that its certificates and reports evidence compliance only with one or more of the

Rules, Guides, standards, or other criteria of Class in accordance with the terms of such certificate or report. Under no circumstances whatsoever should these representations be deemed to relate to any third party.

Definitions

Assembly Secured Shackle. A shackle fitted to a sling leg and secured by a seal or similar device, so as to signal unambiguously, whether or not the shackle has been exchanged.

Breaking Force. The maximum load at which a tensile failure occurs in the sample of wire rope being tested.

Corner Fitting. A corner fitting is a fixture consisting of standard apertures and faces which provide a common interface for handling and securing containers.

Designer. The person or legal entity that has proprietary rights to the design and its use. This may be the OEM.

Lifting Set. Items of integrated equipment used to connect the offshore container to the lifting appliance. This can comprise one or multi-leg slings (with or without a top leg) and shackles, whether assembly secured or not.

Non-conformance. Non-fulfilment of a specified requirement.

Offshore Container. A portable unit with gross mass not exceeding 25,000 kg (55,116 lb), for repeated use in the transport of goods or equipment handled in open seas to, from, and between fixed and/or floating installations and ships. *Note: The unit incorporates permanently installed equipment for lifting and handling and may include equipment for filling, emptying, cooling, heating, etc.*

Original Equipment Manufacturer (OEM). The person or legal entity that has the legal or patent rights to produce the material, component, product, or system.

Owner. Legal owner of the offshore container or the delegated nominee of that body.

Permanent Equipment. Equipment that is attached to the container and which is not cargo. *Note: This may include lifting sets, refrigeration units, shelves, securing points, rubbish (US: garbage) compactors.*

Primary Structure. Load carrying and supporting frames, load carrying panels, supporting structures for tanks, pad eyes, etc. Primary structure is divided into the following two subgroups:

Essential/Non-redundant Primary Structure

Main structural elements which transfer the cargo load to the crane hook (i.e., forming the "load path" from the payload to the lifting sling) include but are not limited to:

- Top and bottom side rails
- Top and bottom end rails
- Corner posts
- Pad eyes; and
- Forklift pockets.

Other primary structures may also be considered as essential/non-redundant.

Non-essential Primary Structure. Structural elements such as floor plates, protective frame members, etc. which do not have their functional requirement are to be categorised here. Side and roof panels (including corrugated panels) are not considered to be part of the primary structure and are not taken into account when evaluating the strength of the container.

Production Units. Production units are identical containers built under conditions which duplicate, as far as is practicable, the conditions under which the prototype was built.

Proof Load. The specific tension applied to a sling or component in the performance of a proof test

Proof Test. A non-destructive tension test of the sling or components.

Prototype. Prototype is a representative unit of a series of identical containers built under conditions which duplicate, as far as is practicable, the conditions under which all of the containers in the series are to be fitted.

Prototype Testing. This is the destructive and non-destructive testing of the materials and components presented for evaluation of the original design of a product. If a Surveyor's witness is required, this may not be waived under any sections of the Rules, unless it is done by a recognised third party.

Secondary Structure. Parts which are not considered as load carrying for the purpose of the design calculations, including the following components:

- Doors, wall, and roof panels
- Panels stiffeners and corrugations
- Structural components used for tank protection only; and
- Internal securing points.

Note: *Not all container walls are corrugated.*

Symbols

F_S	Resulting sling force
g	Acceleration due to gravity
ℓ_n	Nominal reference length (often different from actual span of a beam)
n	Number of pad eyes (for calculation purposes, n is to be either 2 or 4)
P	Payload (i.e., the maximum permissible mass of cargo which may be transported by the container)
R	Rating (i.e., the maximum gross mass of the container including permanent equipment and its cargo, but excluding the lifting set)
S	Mass of the lifting set
SF	Safety Factor
T	Tare mass (i.e., the mass of an empty container including any permanent equipment but excluding cargo and lifting set)
t	Material thickness
t_{min}	Minimum material thickness
T_D	Design air temperature (i.e., a minimum reference temperature used for the selection of steel grades used in offshore containers and equipment expressed in degrees centigrade)
WLL_{min}	Working load limit in tonnes (i.e., the maximum amount of mass that a lifting component is authorised to sustain in lifting service)
WLL_s	Minimum working load limit for each shackle/sling in tonnes
y	Deflection of structural member
β	Angle of sling leg from vertical in degrees
σ_e	Allowable Von Mises equivalent stress
σ_y	Specified minimum yield stress
σ_u	Specified minimum tensile strength at room temperature
ε	0.2% proof stress at room temperature
ψ	Dynamic factor ($= 3$)

Note:

1. $P = R—T$
2. *The term "Safe Working Load" is never used in this guide as it is not clearly defined for containers and for the same reason it is not used when referring to offshore containers. The term "Working Load Limit" is only used for lifting sets and not for containers.*

Contents

1 General Provisions and Scope 1
   1.1   General 1
   1.2   Relationship with Other Standards, Codes and Regulations 2
          1.2.1   IMO-MSC/Circ.860 2
          1.2.2   ISO 10855:2018 2
          1.2.3   EN 12079:2006 3
          1.2.4   IMDG, US DOT, RID/ADR 3
          1.2.5   Dropped Object Prevention on Offshore Units
                  and Installations 3

2 Certification Procedure 5
   2.1   Certification Procedure for Offshore Containers 5
          2.1.1   Application for Certification 5
          2.1.2   Design Review 6
          2.1.3   Quality Control 6
          2.1.4   Production 7
          2.1.5   Class Production Certificate 7
          2.1.6   Certification to Other Standards 7
   2.2   Certification Procedure for the Lifting Set 8
          2.2.1   Components 8
          2.2.2   Application for Certification 8
          2.2.3   Design Review 9
          2.2.4   Quality Control 9
          2.2.5   Production 10
          2.2.6   Class Production Certificate 10

3 Quality Assessment 11
   3.1 General 11
   3.2 Quality Manual 11
   3.3 Quality Plan 12
   3.4 Quality Assessment 13
      3.4.1 Initial Assessment 13
      3.4.2 Annual Assessment 14
      3.4.3 Renewal Assessment 14
      3.4.4 Quality Assessment Report 14
      3.4.5 Overdue Assessment 15
   3.5 Factory Approval Certificate 15
      3.5.1 Certificate 15
      3.5.2 Validity 16

4 Offshore Container Design 17
   4.1 General 17
   4.2 Offshore Containers 17
      4.2.1 Offshore Freight Container 17
      4.2.2 Offshore Service Container 18
      4.2.3 Offshore Waste Skip 18

5 Materials and Welding 21
   5.1 General 21
   5.2 Identification of Materials 21
   5.3 Steels 21
      5.3.1 Toughness Requirements 22
   5.4 Steel Casting in ISO-Corner Fittings 22
   5.5 Aluminium 22
   5.6 Non-metallic Materials 23
   5.7 Material Certificates 24
   5.8 Welding 24
      5.8.1 Welding Procedure 24
      5.8.2 Welding of Primary Structure 25
      5.8.3 Welding of Secondary Structure 25
      5.8.4 Welder/Welder Operator Qualification 25

6 Design 27
   6.1 General 27
      6.1.1 Structural Design 27
      6.1.2 Stability from Overturning 27
      6.1.3 Protection for Protruding Parts 28
      6.1.4 Top Protection 28

		6.1.5	Intermediate Cargo Decks	29
		6.1.6	Design Temperature (T_D)	29
		6.1.7	Stacking and Stacking Fittings	29
	6.2	Structural Strength		29
		6.2.1	Lifting with Lifting Set	30
		6.2.2	Lifting with Forklift Truck	30
		6.2.3	Impact Loads	31
		6.2.4	Pad Eye Design	32
		6.2.5	Internal Forces on the Container Walls	33
	6.3	Other Structural Requirements and Construction		33
		6.3.1	Primary Structure	33
		6.3.2	Secondary Structure	34
		6.3.3	Additional Structure	35
	6.4	Equipment		36
		6.4.1	Coating and Corrosion Protection	36
7	**Prototype Testing**			37
	7.1	Test Equipment and Calibration		37
	7.2	Prototype Testing		37
	7.3	Lifting Test		38
		7.3.1	General	38
		7.3.2	All-Point Lifting	38
		7.3.3	Two-Point Lifting	38
		7.3.4	Post-lifting Test Inspection and Examination of Pad Eye	38
	7.4	Vertical Impact Test		39
		7.4.1	General	39
	7.5	Other Tests		39
		7.5.1	Forklift Pockets	39
		7.5.2	Driving Ramps	40
		7.5.3	Stability from Overturning	40
	7.6	Dangerous Goods Cargo		40
8	**Production and Manufacture of Offshore Containers**			41
	8.1	General		41
	8.2	Primary Structure		42
		8.2.1	Examination of Welds	42
	8.3	Secondary Structure		43
	8.4	Production Testing		43
		8.4.1	Lifting Test	43
		8.4.2	Weather Proofness Testing	44

9   **Offshore Container Marking and Data Plates** 45
    9.1   Safety Marking 45
    9.2   Identification Markings 45
           9.2.1   Container Number 45
    9.3   Information Marking 45
    9.4   Marking for Containers with an Intermediate Deck 47
    9.5   Container Data Plate 47
           9.5.1   General 47
           9.5.2   Contents on the Data Plate 47

10 **Lifting Sets** 49
    10.1   General 49
    10.2   Materials and Welding 49
           10.2.1   General 49
           10.2.2   Identification of Materials 49
    10.3   Materials Used in Slings and Their Components 50
    10.4   Toughness Requirements 50
    10.5   Galvanising 50
    10.6   Material Certificates 50
    10.7   Welding 50
    10.8   Technical Requirements 51
           10.8.1   General Requirements 51
           10.8.2   Dimensions and Strength of Lifting Sets 51
    10.9   Components 51
           10.9.1   Wire Rope Slings 51
           10.9.2   Chain Slings 53
    10.10   Master (Top) Link 53
           10.10.1   Ferrules 53
           10.10.2   Thimbles 54
           10.10.3   Shackle 54
    10.11   Marking 55
           10.11.1   Shackles 55
           10.11.2   Slings 55

11 **In-Service Inspections** 59
    11.1   General 59
    11.2   Schedule of Inspection for Offshore Containers 59
           11.2.1   Schedule of Periodic Inspection, Examination
                      and Test—Offshore Containers 60
           11.2.2   Lifting Test 60

	11.2.3	Non-destructive Examinations	60
	11.2.4	Visual Inspection	60
11.3	Schedule of Inspection, Examination and Tests—Lifting Sets		61
	11.3.1	Load Test of Chain Sling Legs	61
	11.3.2	Non-destructive Examination of Sling Components Except Wire Rope Legs	62
	11.3.3	Visual Inspection of the Lifting Set	62
11.4	Inspection Plates/Tags		62
	11.4.1	Offshore Containers Inspection Plate	62
	11.4.2	Contents of the Inspection Plate	63
	11.4.3	Marking	63
11.5	Inspection Tag for Lifting Set		64
	11.5.1	Marking	64
11.6	Existing Offshore Containers		64
	11.6.1	General	64
	11.6.2	Marking and Date Plate	65
11.7	Emblem		66

12 Repair and Modification Procedures ... 69
12.1	General	69
12.2	Modification Procedures	70
12.3	Recordkeeping	70

13 Class Recognised Standards ... 71
13.1	General	71
13.2	Materials	71
13.3	Quality Control for Manufacturing Facilities	72
13.4	Iso Corner Fittings	72
13.5	Padeyes	72
13.6	Welding	72
13.7	Calibration	73
13.8	Non-destructive Testing	73
13.9	Weatherproofness Testing	74
13.10	Tank Containers	74
13.11	Containers for Bulk Solids	74
13.12	Hazardous Areas	74
13.13	Analysis—Beam Elements	74
13.14	Standards for Lifting Set and Their Components	75

14 Certification Requirements for Offshore Tank Containers/Portable Tanks and MEGC ... 77
| 14.1 | Tank Containers for Dangerous Goods | 77 |
| | 14.1.1 | Tank Containers—General | 77 |

         14.1.2   Frame 77
         14.1.3   Tanks for Fluids 77
         14.1.4   Impact Protection on Tank Containers for Dangerous
                  Cargoes 78
         14.1.5   Periodic Inspections for Tank Containers 78
         14.1.6   Multiple Element Gas Containers (MEGCs) 79

15   Container Corner Castings 81
     15.1   General 81
     15.2   Process of Manufacture 81
     15.3   Heat Treatment 81
     15.4   Material Specifications 82
            15.4.1   Chemical Analysis 82
            15.4.2   Chemical Requirements 82
     15.5   Tension Test 82
            15.5.1   Tensile Test Specimen 83
            15.5.2   Retests 83
            15.5.3   Tensile Properties 83
            15.5.4   Charpy Impact Test 83
     15.6   Inspections 84
            15.6.1   Dimensional Inspection 84
            15.6.2   Visual Inspection 84
            15.6.3   Internal Discontinuities Examination 84
     15.7   Marking 85

Correction to: Guidance on the Classification of Offshore Containers C1

Annex 87

Abbreviations and Acronyms

AA	American Aluminium Association
ADR	International Carriage of Dangerous Goods by Road
AISC	American Institute of Steel Construction
ASME	American Society of Mechanical Engineers
ASTM	American Society for Testing and Materials
BL	Breaking Load
CSC	International Convention for Safe Containers
CVN	Charpy V-Notch
IACS	International Association of Classification Societies
IMDG	International Maritime Dangerous Goods Code
IMO	International Maritime Organisation
ISO	International Standards Organisation
MEGC	Multiple Element Gas Containers
MTR (1)	Material Traceability Report
MTR (2)	Material Test Report
NDE	Non-destructive Examination
NDT	Non-destructive Testing
PL	Proof Load
PQR	Procedure Qualification Record (PQR)
RID	International Carriage of Dangerous Goods by Rail
T_D	Design Air Temperature
US DOT	United States Department of Transport
WLL	Working Load Limit
WLL_{min}	Minimum Working Load Limit
WPQR	Welder Performance Qualification Records (WPQR)
WPS	Welding Procedure Specification (WPS)

List of Figures

Fig. 4.1   Offshore cargo container 19
Fig. 4.2   Suspended offshore cargo container above deck 19
Fig. 9.1   Example of data plate layout 48
Fig. 10.1   Construction of wire rope 52
Fig. 10.2   Four-legged wire sling with fore runner 52
Fig. 10.3   Chain slings 53
Fig. 10.4   Master link assembly 54
Fig. 10.5   Thimble 54
Fig. 10.6   Shackles 55
Fig. 10.7   Identification tag for chain sling 56
Fig. 10.8   Identification tag for wire rope sling 57
Fig. 11.1   Example of inspection plate layout 65
Fig. 11.2   Template of combined data plate and inspection plate 66

List of Tables

Table 5.1   Charpy impact test—structural steel for primary structural members 22

Table 5.2   Aluminium alloys and tempers for rolled products 23

Table 5.3   Aluminium alloys and tempers for extruded products 23

Table 5.4   Documentation of materials 24

Table 8.1   Non-destructive examination (NDE) of structural welds 42

Table 8.2   Standards relevant (NDE) methods 42

Table 8.3   NDE Acceptance criteria 43

Table 8.4   Number of containers required for lifting test 43

Table 9.1   Safety marking 46

Table 9.2   Container number 46

Table 10.1   Required minimum shackle working load limit (WLLs) 51

Table 11.1   Schedule of inspection, examination and tests for offshore containers 60

Table 11.2   Schedule of periodic inspection, examination and testing of lifting sets 62

Table 15.1   Chemical composition (%) 82

General Provisions and Scope

1

1.1 General

This guide has been written to provide guidance relating to the requirements for the certification of offshore containers with a gross mass not exceeding 25,000 kg (55,116 lb), intended for repeated use to, from and between fixed or floating installations and ships. These containers may not always conform to the standard design requirements of the International Convention for Safe Containers (CSC). Accordingly, some of the features that differentiate offshore containers from the containers that are designed according to the CSC are indicated below:

- Containers conforming to CSC convention are designed for loading and unloading in ports and in inland transport only and are not designed for handling in the open seas. However, offshore containers are designed to be lifted onto and from offshore installations and ships, and such operations may often take place in unfavourable weather and sea conditions
- Offshore containers are not intended to be lifted using spreader beams or ISO Corner fittings and are to be lifted using pad eyes or lifting lugs with the designated lifting sets only; and
- Offshore containers, unlike ISO containers, are not standardised with respect to sizes or gross mass.

Therefore, the intention of this guide is to provide an outline standard for containers that are to be utilised in the offshore industry through conformity with the guidelines issued by the IMO in MSC/Circ.860. This guide only applies to the transport and lifting requirements for offshore containers with respect to design, manufacturing, marking, testing and periodic inspections. This guide does not include requirements for containers intended for intermodal transport. For this application refer to the specific Class Rules for the certification of cargo containers. Furthermore, this guide does not include requirements for the temporary or permanent installation of offshore containers on fixed or floating offshore facilities. For these applications refer to the Class Rules for portable accommodation modules. Offshore containers that are certified to the requirements of this guide will typically comply with MSC/Circ.860, EN 12079:2006 and ISO 10855:2018. Offshore Container Production Certificates issued by Class will indicate this compliance where appropriate. Certification of existing containers may be considered by Class on a case-by-case basis. Further guidance on this matter is provided in Chap. 14.

1.2   Relationship with Other Standards, Codes and Regulations

This guide references standards recognised by the majority of classification societies who are members of the International Association of Classification Societies (IACS), which are listed in Chap. 14.

1.2.1   IMO-MSC/Circ.860

As mentioned above, the IMO has issued guidelines for the approval of offshore containers, in circular MSC/Circ.860. This circular is intended to guide national authorities in developing approval and certification requirements for offshore containers. It recommends that new offshore containers be approved, prototype tested, and certified by duly authorised bodies (including, but not necessarily limited to, Class).

1.2.2   ISO 10855:2018

ISO 10855:2018 is one of the standards used by the offshore industry for the certification of offshore containers and associated lifting sets. This standard consists of three parts as described below:

- EN 10855:2018, Part 1: Design, Manufacturing and Marking of Offshore Containers
- EN 10855:2018, Part 2: Manufacturing and Marking of Lifting Set; and
- EN 10855:2018, Part 3: Periodic Inspection, Examination and Testing.

1.2.3 EN 12079:2006

EN 12079:2006 is a standard used in the offshore industry for the certification of offshore containers and associated lifting sets. This standard consists of three parts as described below:

- EN 12079:2006, Part 1: Offshore Containers—Design, Manufacturing and Marking
- EN 12079:2006, Part 2: Lifting sets—Design, Manufacturing and Marking; and
- EN 12079:2006, Part 3: Periodic Inspection, Examination and Testing.

1.2.4 IMDG, US DOT, RID/ADR

Offshore containers that intended for the carriage of dangerous cargo are required to comply with the International Maritime Dangerous Goods (IMDG) Code. In any number of cases, offshore containers may also be required to conform to other national and international standards and regulations such as United States Department of Transport (US DOT), Regulations concerning the International Carriage of Dangerous Goods by Rail (RID) and the European Agreement concerning the International Carriage of Dangerous Goods by Road (ADR), etc.

1.2.5 Dropped Object Prevention on Offshore Units and Installations

Reference may be made to *Dropped Object Prevention on Offshore Units and Installations*, also published by Springer, to provide authoritative guidance with reference to dropped prevention programmes. These correspond to a growing industry-wide move towards mitigating and ultimately eliminating the hazards imposed by dropped objects. The application of guidance from this publication to offshore containers is optional. In any case, offshore containers may be reviewed by Class for compliance with Class standards for the prevention of dropped objects on offshore units and installations at the request of the designer.

Certification Procedure

2

2.1 Certification Procedure for Offshore Containers

The certification procedure outlined below is a typical procedure which will facilitate the certification of offshore containers in accordance with the guidance outlined in this guide. That said, Class may consider alternative procedures, provided they are no less effective than the requirements indicated in this guide.

2.1.1 Application for Certification

An application is to be submitted by the designer or manufacturer who has the legal right to the design. The designer and manufacturer may not be the same entity, in which case this must be clearly stated on the application. The application is to be submitted with, but not limited to, the following information and data:

(1) Design drawings including, but not limited to, the following information:
  - Dimensions
  - Maximum gross mass and payload
  - Scantling of members
  - Material properties
  - Pad eye details
  - Welding details
  - Markings
(2) Design calculations
(3) Testing results (if applicable); and
(4) Declaration for the absence of asbestos.

© The Author(s), under exclusive license to Springer Nature Switzerland AG 2025   5
A. A. Olsen, *Guidance on the Classification of Offshore Containers*, Synthesis Lectures on Ocean Systems Engineering, https://doi.org/10.1007/978-3-031-74857-8_2

Applications for additional units to be certified under an approved design are to include at least the application and marking drawings if the owner has changed them. Revisions to an existing Class approved design may require an application which is to be submitted with, but not limited to, the applicable documents indicated above. Submitted documents are to completely describe the changes therein.

2.1.2   Design Review

In most cases, the documents that are submitted with the application will be reviewed by a Class engineer. The engineer will evaluate the design under the Class requirements as well as any other relevant and appropriate Rules, regulations, and/or standards. Upon a successful review, a Class review letter with test agenda, if applicable, will be returned indicating that the design meets the requirements of Class. The prototype test agenda is usually provided only after all review items have been addressed. Importantly, the design is not considered Class approved until the prototype test is conducted in the presence of a Class Surveyor in accordance with the prototype test agenda specified by Class.

2.1.3   Quality Control

The manufacturing facility for the offshore container is required to submit a quality manual to the Class Surveyor, in which it describes the quality system. The Surveyor will review the manual to verify all aspects of the production of the container are addressed in the manual. Following a successful review of the manual, a quality audit may be performed at the manufacturing facility to verify that the system outlined in the manual is in place and functioning properly. Upon a successful quality audit, the Class Surveyor will issue a Factory Approval Certificate. These certificates are valid for five years, with annual endorsements required.

Notes:

(1) *A quality control document for the manufacturing facility is required to be submitted only if the facility is submitting for the first design to be manufactured at the facility.*
(2) *Any updates to the quality control documents are to be submitted for review. The document is to be updated if a new design requiring additional quality procedures is to be produced at the facility.*

2.1.4   Production

Upon a successful design review, prototype test and quality assessment, the manufacturing facility may begin production under the surveillance of the Class Surveyor. The Class Surveyor will initially request a meeting to discuss inspection schedules, hold points, and other related items. Welding Procedure Specifications (WPS) and Welder Qualifications will almost certainly be reviewed and subject to approval by the Class Surveyor prior to the commencement of production. Production testing is to be carried out in the presence of a Class Surveyor or quality representative nominated by the manufacturer as indicated in the Class approved quality document. The Surveyor will "walk the production line" during each attendance to verify that the quality system continues to function properly. The below documents are to be presented to the Class Surveyor at the time of final inspection in an as-built dossier to support the certification of the offshore container. The documents must include, but are not necessarily limited to, the following:

(1) Material certificates
(2) Material Traceability Reports (MTR) for primary structure
(3) Fabrication inspection reports
(4) Dimensional control reports
(5) Non-destructive Testing (NDT) reports
(6) Production testing reports; and
(7) Pressure testing records.

2.1.5   Class Production Certificate

Following a successful review of the as-built dossier and final inspection, a Class Production Certificate will be issued by the Class Surveyor. Additionally, the certificate will reference the Class engineering approval letter and prototype test certificate, applicable to the container design.

2.1.6   Certification to Other Standards

When the application includes a request for certification to governmental requirements, international conventions, or other standards, the submission is to include the necessary required information in totem. These additional requirements will be indicated on the certificate.

2.2   Certification Procedure for the Lifting Set

With reference to lifting sets, the certification procedure outlined below is a typical proce-dure which will facilitate the certification of the lifting sets. Class may consider alternative procedures, provided they are no less effective than the requirements outlined in this guide. Lifting sets must be assembled from the various components described within this guide. The components are to meet the requirements of Class and applicable standards and be presented to the manufacturer of the lifting set with supporting documentation. The lifting set will be certified in the assembled condition by the attending Class Surveyor.

2.2.1   Components

Manufacturers of the following components are to submit the component for certification through the Class Type Approval programme:

- Shackles
- Chains
- Links (master links and master link assemblies, intermediate links, end links); and
- Couplings.

The Class Type Approval procedure as well as the application can typically be found on the Class website or on request. The above components should be presented to the lifting set manufacturer with traceability to the Class Type Approval certificate. Other components such as wire rope, ferrules, and thimbles do not require Class Type Approval. However, they are to be presented to the lifting set manufacturer with a 3.1 Certificate in accordance with ISO 10474 or another standard recognised by Class.

2.2.2   Application for Certification

An application is to be submitted by the designer or manufacturer who has the legal right to the design of the lifting set. The designer and manufacturer may not be the same entity, in which case this is to be clearly stated on the application. Documents to be submitted with the application include, but are not necessarily limited to, the following:

(1) Specification of the lifting set, along with the following:
   - Applicable standards
   - Material specification
   - Dimensions of all the components
   - Working Load Limit (WLL)

- Proof Load (PL)
- Breaking Load (BL)

(2) Drawings of lifting set and its components
(3) Calculations demonstrating how the lifting set components were selected, including determination of required strength
(4) Description of all the manufacturing/assembly procedures (e.g., for wire rope sling: the assembly of the sling legs with terminal, etc.)
(5) Marking details on the tags
(6) Description of the test methods and procedures for all relevant prototype and production tests; and
(7) Type approval/3.1 certificates for the components according to Chap. 1, Sect. 2.1.

Applications for additional units to be certified under an approved design are to include at least the application and marking drawings if the owner has changed them. Revisions to an existing Class approved design will require an application which is to be submitted with, but not limited to, the applicable documents indicated above. Submitted documents are to completely describe the changes.

2.2.3   Design Review

The documents submitted with the application will be reviewed by a Class engineer. The engineer will evaluate the design under the requirements set by Class in addition to any other requested Rules, regulations or standards. Upon a successful review, a Class review letter with test agenda, if applicable, will be returned indicating that the design meets the requirements set by Class. The prototype test agenda is only provided after all review items are addressed. The design is not approved until the prototype test is conducted in the presence of a Class Surveyor in accordance with the test agenda specified by Class.

2.2.4   Quality Control

The manufacturing facility is required to submit a quality manual to the Class Surveyor, in which it describes the quality system. The Surveyor will review the manual to verify all aspects of the production of the lifting set are addressed in the manual. Following a successful review of the manual, a quality audit may be performed at the manufacturing facility to verify that the system outlined in the manual is in place and functioning properly. Upon a successful quality audit, the Class Surveyor will issue a Factory Approval Certificate. These certificates are valid for five years, with annual endorsements required.

Notes:

(1) *A quality control document for the manufacturing facility is required to be submitted only if the facility is submitting for the first design to be manufactured at the facility.*
(2) *Any updates to the quality control documents are to be submitted for review. The document is to be updated if a new design requiring additional quality procedures is to be produced at the facility.*

2.2.5   Production

Upon a successful design review, prototype testing and quality assessment, the manufacturing facility may begin production of lifting sets in accordance with the approved design. Testing carried out during the production phase are to be in accordance with the relevant sling or component standards.

2.2.6   Class Production Certificate

A Class Production Certificate will be issued by the Surveyor following a successful review of the as-built dossier (which includes the Class Type Approval or 3.1 certificates for the components of the lifting set) and final inspection. Additionally, the certificate will reference the Class engineering approval letter and prototype test certificate, applicable to the lifting set design.

Quality Assessment

3

3.1 General

In general, the quality assessment will consist of a review of the quality manual and audit of the facility to verify manufacturing is performed in accordance with the manual. Typically, the quality assessment will consist of:

(1) Contract or application
(2) Quality Manual review
(3) Quality Plan review
(4) Management assessment
(5) Production assessment; and
(6) Certification.

3.2 Quality Manual

The purpose of the quality manual is to describe the scope and extent of the company's quality system in a concise and brief format. Therefore, the company is expected to establish and maintain a quality manual that includes, but is not limited to, the following:

(1) The scope of quality management system, including details of and justification for any exclusions
(2) The documented procedures established for the quality management system, or reference to them; and
(3) A description of the interaction between the various processes within the quality management system.

© The Author(s), under exclusive license to Springer Nature Switzerland AG 2025   11
A. A. Olsen, *Guidance on the Classification of Offshore Containers*, Synthesis Lectures on Ocean Systems Engineering, https://doi.org/10.1007/978-3-031-74857-8_3

The quality manual to be submitted to Class for review and approval prior to lodging a request for an initial assessment. However, where a recognised certification body has approved the quality manual, Class may exercise their discretion to not require the manual to be submitted to Class for review. In all circumstances, the quality manual is to be available for the Class Surveyor to assess the performance of the quality system in place at the manufacturing facility.

3.3   Quality Plan

A typical quality plan is used to describe the methods of assuring and controlling quality during production as may be required by the product specifications and will be subject to review by Class. In particular, the quality plan is to reflect specific inspections, tests, etc., required by Class Rules and Guides, as well as international regulations and standards. The manufacturer is to present a representative sample of the product "type" to the Surveyor for the purpose of verifying that the "type" has been manufactured in conformance with the design documents. Prior to the assessment, the manufacturer is to submit the quality plan to the Surveyor. The plan is to include, but may not be limited to, the following:

- Issuance of material specification for purchasing
- Receiving inspection of materials
- Receiving inspection of finished components and parts
- Calibration certification
- Dimensional and functional checks on finished components and parts
- Edge preparation and fit-up tolerances
- Welding procedure qualifications
- Welder qualification
- Welding defect tracking
- Non-destructive test written procedures and qualification documentation
- Non-destructive test plan
- Casting and weld defect resolutions
- Assembly and fit specifications
- Subassembly inspection: alignment and dimensional checks, functional tests
- Testing of safety devices
- Hydrostatic testing plan
- Factory acceptance test plan; and
- Identification of dropped object prevention features (if applicable).

3.4   Quality Assessment

An assessment is a systematic and independent examination to determine whether quality, environmental, financial, other management activities and the related results comply with planned arrangements, and whether these arrangements are being implemented effectively, and are suitable to achieve the assurance of quality standards. To that end, there are two categories of quality assessment:

(1) *Management Assessment.* These evaluate the quality assurance and quality control system of the manufacturing facility to verify its capability to consistently meet the manufacturer's specified level of product quality and satisfy the requirements of the Class Rules and Guides, as well as international regulations and/or standards; and
(2) *Production Assessment.* These evaluate the product specific manufacturing process in order to verify that the manufacture and inspections of the products are established to meet the manufacturer's specified level of quality control and, to satisfy the requirements of the Class Rules and Guides, as well as international regulations and/ or standards.

Items that are periodically renewed which require verification by the Class Surveyor are to be obtained, verified, and attached to the assessment report.

3.4.1   Initial Assessment

For an initial assessment, the manufacturer must submit an application to Class requesting an initial assessment. The application should include a copy of the manufacturer's quality manual, which will be reviewed by Class prior to the initial assessment at the manufacturer's location. It bis worth noting the quality assurance system is more comprehensive than the manufacturing process since it considers all the factors that affect and impact on the process. Therefore, the system includes, but is not necessarily limited to, any and all of the following:

- Design Assessment
- Quality Manual
- Quality Plan
- Control of process inputs
- Process controlling factors (e.g., competency of personnel, procedures, facilities and equipment, training, etc.)
- Process outputs
- Measurements of quality

- Process and product for continual improvement; and
- Control of contracted vendors, service providers, and suppliers.

3.4.2   Annual Assessment

The manufacturer must be able to produce records of the products continued compliance with the standard. Calibration certificates for each piece of equipment used in the production of the container are to be collected and collated during the annual quality assessment and thereafter retained as part of the endorsement.

3.4.3   Renewal Assessment

The manufacturer must submit an application to Class for the renewal of an existing quality assessment at least 90 days prior to the expiration date of the current quality Factory Approval Certificate. Where for a practical reasons the renewal process of the Factory Approval Certificate cannot be completed before the expiration date of the current certificate, a short-term extension may be considered by Class upon application. Where the certificate is renewed within 90 days of the expiration date, the new certificate will be valid for five years from the expiration date of the previous certificate. The renewal assessment is to be no less detailed than an initial or annual assessment. During the renewal process, the Class Surveyor will typically verify that:

- There have been no changes to the design
- The design assessment indicates the most current Class Rules and Guides, and/or international regulations and standards (as may be applicable); and
- The quality plan remains effective to control quality during production.

3.4.4   Quality Assessment Report

On receipt of the quality assessment report, the applicant is expected to acknowledge any comments or observations. The applicant is required to take corrective action on all instances of non-conformance and to remedy and conditions found. Corrective actions are to be clearly detailed in the auditor's report.

3.4.4.1   Findings, Non-conformances, and Observations

There are subtle differences between findings, non-conformances, and observations. These are briefly outlined below:

(1) *Finding.* A statement of fact supported by objective evidence about a process whose performance characteristics meet the definition of non-conformance or observation
(2) *Non-conformance.* A non-conformance is the identification of a non-fulfilment of a specified requirement; and
(3) *Observation.* An observation is a statement of fact made during a system audit and substantiated by objective evidence. It may also be a statement made by the auditor referring to a situation within the Management System which, if not corrected, may lead to a nonconformity in the future. Therefore, in all subsequent audits, previous observations are to be reviewed to determine if they have become non-conformities.

Initial, annual or renewal certification will not generally be credited wherever a non-conformance is identified. Unless stipulated otherwise by Class, non-conformances found at initial, annual or renewal assessments must be addressed within 90 days of the audit.

3.4.5 Overdue Assessment

If an annual or renewal audit is not completed within 90 days after the anniversary date of the Factory Approval Certificate, all production work is to be inspected for verification of compliance with the latest Class Rules and Guides, and/or international regulations and/or standards (as may be applicable).

3.5 Factory Approval Certificate

The approval of the factory is based on the assessment outlined above.

3.5.1 Certificate

Manufacturing facilities will be issued a Factory Approval Certificate once they have been audited and found to comply with the following requirements:

(1) Have undergone a satisfactory design evaluation
(2) Comply with a quality assurance standard; and
(3) Have manufacturing quality control that meets the applicable provisions of the Rules, product standard, or manufacturer's specification.

3.5.2   Validity

Each Factory Approval Certificate is valid for five years and is subject to annual endorsements. In most cases Class may reserve the right to perform unscheduled assessments without notice. The Factory Approval certificate is not transferable and is issued to a unique manufacturer, at a specific address, with specific ownership and a specific organisation.

Offshore Container Design

4

4.1 General

This chapter specifies the requirements for the design, manufacture and marking of offshore freight, waste skip and service containers intended for repeated use to, from and between fixed or floating installations and ships.

4.2 Offshore Containers

An offshore container is a portable unit with gross mass not exceeding 25,000 kg (55,116 lb) for repeated use in the transport of goods or equipment handled in open seas to, from and between fixed and/or floating installations and ships. Offshore containers are generally subdivided into the following three categories.

4.2.1 Offshore Freight Container

Offshore freight containers are offshore containers built for the transport of goods. Examples of offshore containers include:

- General Cargo Container. A closed container with doors
- Cargo Basket. An open top container for general or special cargo
- Tank Container. A container for transport of dangerous or non-dangerous fluids
- Multiple Element Gas Containers (MEGCs)
- Bulk Container. A container for the transport of solids in bulk

© The Author(s), under exclusive license to Springer Nature Switzerland AG 2025   17
A. A. Olsen, *Guidance on the Classification of Offshore Containers*, Synthesis Lectures on Ocean Systems Engineering, https://doi.org/10.1007/978-3-031-74857-8_4

- Special Container. A container for the transport of special cargo (e.g., rubbish (US: garbage) containers, equipment); and
- Boxes, gas cylinder racks.

4.2.2 Offshore Service Container

An offshore service container is an offshore container built and equipped for a special service task, usually as a temporary installation (e.g., laboratories, workshops, stores, power plants, control stations). Offshore service containers on a fixed or floating off-shore facilities are to be in accordance with the specific Class guide relating to portable accommodation modules. In most cases portable accommodations are considered offshore service containers for the purposes of transit only.

4.2.3 Offshore Waste Skip

An offshore waste skip is an open or closed offshore container used for the storage and removal of waste.

Note: *Normally constructed from flat steel plate forming the load bearing sections of the container, with bracing in the form of steel profiles (e.g., channel or hollow section) fitted horizontally and/or vertically around sides and ends.*

   In addition to the pad eyes for the lifting set, these containers may have side mounted lugs suitable for use with the lifting equipment mounted on a skip lift vehicle (see Figs. 4.1 and 4.2).

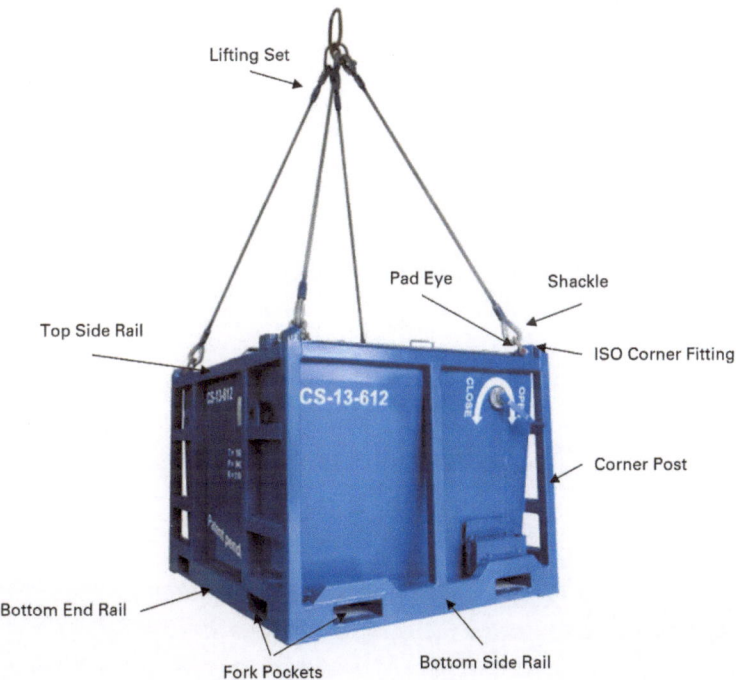

Fig. 4.1 Offshore cargo container

Fig. 4.2 Suspended offshore cargo container above deck

Materials and Welding 5

5.1 General

Materials made of steel must be suitable for the intended service conditions. This means they must be of good quality, free of defects, and are to exhibit satisfactory formability and weldability characteristics. Materials other than steels are to be specially considered during the Class review.

5.2 Identification of Materials

The manufacturer is to adopt a system for the identification of finished plates, shapes, castings, and forgings which will enable the material to be traced to its original heat; moreover, the Surveyor is to be given sufficient documentation, such as a Material Test Report (MTR) and a process for verifying the grade markings and traceability of the material.

5.3 Steels

Structural steels for the primary structure are to be carbon steel, carbon-manganese steel, carbon manganese micro-alloyed steel, or low-alloyed steel. Rolled (plates, profiles, or hollow sections) or forged or extruded or cast steels are to be produced in accordance with a standard recognised by Class. Steels in the primary structure must be killed and fine grain treated. Steels with yield strength (σ_y) above 500 N/mm^2 (51 kgf/mm^2, 73 ksi)

© The Author(s), under exclusive license to Springer Nature Switzerland AG 2025 21
A. A. Olsen, *Guidance on the Classification of Offshore Containers*, Synthesis Lectures on Ocean Systems Engineering, https://doi.org/10.1007/978-3-031-74857-8_5

Table 5.1 Charpy impact test—structural steel for primary structural members

Yield strength			Average CVN (longitudinal)			Test temperature	
N/mm^2	kgf/mm^2	ksi	J	kgf-m	ft-lbs	Thickness ≤25 mm (0.98 in.)	Thickness >25 mm (0.98 in.)
235–305	24–31	34–44	27	2.8	20	T_D	20 °C (36 °F) below T_D
305–420	31–43	44–61	σy/10	σy/1.01	σy/1.97		
420–500	43–51	61–73	42	3.5	31		

Note Steels with thickness <6 mm (0.23 inches) do not require CVN tests

are not to be used. Where required, steels for welding are to be made by open hearth, electric furnace, or the basic oxygen steel process. Stainless steels are to comply with the requirements of a standard recognised by Class.

5.3.1   Toughness Requirements

Charpy V-Notch (CVN) tests are required to demonstrate that rolled or forged or extruded steels would meet the longitudinal CVN impact requirements indicated in Table 5.1. CVN tests are to be carried out in accordance with ISO 148-1 or other Class recognised standards.

5.4   Steel Casting in ISO-Corner Fittings

Refer to Chap. 15 for requirements.

5.5   Aluminium

The chemical composition, heat treatment, weldability and mechanical properties are to be suitable for the purpose. When materials of different galvanic potential are joined together, the design is to be such that galvanic corrosion to be avoided. Aluminium alloys used in offshore containers are to be made by rolling or extruding. Aluminium alloys and tempers as specified in Tables 5.2 and 5.3 may be used. Use of other alloys or tempers are to be subject to special consideration.

Table 5.2 Aluminium alloys and tempers for rolled products

Alloy		Temper
ISO 209-1	AA	ISO/AA
AlMg 2.5	5052	0/0 HAR/H32 HBR/H34 HCR/H36
AlMg 3	5754	0/0 HAR/H32 HBR/H34
AlMg 3.5	5154	0/0 HAR/H32 HBR/H34
AlMg 4	5086	0/0 HAR/H32 HBR/H34
AlMg 3 Mn	5454	0/0 HAR/H32 HBR/H34
AlMg4.5 Mn	5083	0/0 HAR/H32 HBR/H34
AlSiMgMn	6082	0/0 TB/T4 TE/T5 TF/T6

Note AA—American Aluminium Association

Table 5.3 Aluminium alloys and tempers for extruded products

Alloy		Temper
ISO 209-1	AA	ISO/AA
AlSi0.5 Mg	6063	TB/T4 TF/T6
AlSiMgMn	6082	TF/T6

Note AA—American Aluminium Association

5.6 Non-metallic Materials

Timber, plywood, fibre plastics, and other non-metallic materials are not to be used in the construction of primary structures.

Table 5.4 Documentation of materials

Structure	Documentation and certification		
	Inspection certificate 3.2[a]	Inspection certificate 3.1[a]	Test report 2.2[b]
Pad eyes	●		
Other primary structural members		●	
Secondary structural members			●

Notes
[a] Certificate issued equivalent to 3.1 and 3.2 in ISO 10474
[b] Test report is to be equivalent to 2.2 in ISO 10474
[c] Class recognised standard equivalent to ISO 10474 can be used as an alternative

Notes:

(1) *Consideration should be given to strength, durability, suitability, and possible hazards caused by use of these materials.*
(2) *For usage of non-metallic material refer to the Class Guidance Notes on Review and Approval of Novel concepts.*

5.7   Material Certificates

Materials used for the construction of offshore containers are to be furnished with documentation in accordance with Table 5.4. All materials for primary structures are to be identifiable against the certificates.

5.8   Welding

Welding is to be in accordance with a standard recognised by Class as per Chap. 13 or approved manufacturer specifications. Drawings are to indicate the code applied.

5.8.1   Welding Procedure

A written Welding Procedure Specification (WPS) is to be prepared in accordance with a standard recognised by Class. The WPS and the supporting Procedure Qualification Record (PQR) are to be reviewed and accepted by the attending Class Surveyor. Impact tests are required as part of the welding procedure qualification. Test temperatures and test results are to comply with the requirements given Table 5.1. Welding procedures that are not in accordance with a recognised standard are to be submitted for review and approval by the relevant Class materials department.

5.8.2 Welding of Primary Structure

All primary structure is to be welded by full penetration welds. Non-essential primary structure may be joined by fillet welds with the approval of Class. Unwelded primary structures may be accepted by special consideration.

5.8.3 Welding of Secondary Structure

Secondary structure is to be joined by continuous fillet welds including the welding of secondary structure to the primary structure. Intermittent fillet welds may be considered with the approval of Class.

5.8.4 Welder/Welder Operator Qualification

Before proceeding with welding, the welder or the welding operator is to be qualified to the intended welding procedure. Properly documented Welder Performance Qualification Records (WPQR) conducted in accordance with a recognised welding standard (such as the ASME Boiler and Pressure Vessel Code, Section IX, or AWS D1.1) and certified by a recognised body may be presented to the Class Surveyor for acceptance as evidence of qualification. A qualified welder or welding operator is permitted to perform similar welding, provided the welding essential variables (e.g., position, with or without backing, pipe size, etc.) are within specified ranges defined by the recognised welding standard being applied.

Design

6

6.1 General

All offshore containers are to be designed and constructed such that they can withstand impact loads resulting from heavy seas or contact with any structure during transportation. The design is to facilitate loading and unloading operations while the vessel is operating at a maximum wave height of 6 m (20 ft). Suitable means are to be taken to avoid direct contact of faying surfaces of aluminium to steel. The designer is responsible for designing the offshore container with sufficient strength to withstand the design loads/ testing loads and is to include factors of safety allowing for fatigue, normal wear and tear, manufacturing fabrication techniques, and material properties.

6.1.1 Structural Design

Only the primary structure as a structural frame is to be considered in design calculations. Special consideration to the definition of primary structure may be given to certain types of containers (e.g., waste skip). Structural design is to be performed in accordance with this chapter, Sect. 6.2. Transition is to be provided for structural continuity to reduce stress concentrations.

6.1.2 Stability from Overturning

Offshore containers are to be designed to withstand an incline of 30° in any direction loaded at its maximum gross mass applied at its centre of gravity. If the actual centre of gravity is not known, the maximum gross mass is to be applied at the half height of the container. Calculations are to be provided verifying the inclining requirement, alternatively an incline test is to be performed.

6.1.3   Protection for Protruding Parts

Protruding parts are to be clear from damaging other containers and lifting sets during operations. Pad eyes are to be designed in such a way that they do not protrude outside the boundaries of the container other than vertically upward, and as far as possible they are to be designed to avoid damage from other containers.

6.1.4   Top Protection

Open top containers with permanently installed fixtures or equipment and open frame containers are to be provided with top protection made from a robust material (e.g., plates, grating, GRP, tarpaulin, nets/mesh, webbing). Top protections are not to be located lower than the lower flange of the top frame members. Fixtures for the top protection are not to cause snagging hazards. Top protections can be rigid or flexible. Where possible, the top protection shall cover entire roof of the container; small openings may be incorporated to permit the passage of slings when pad eyes are located below the protection.

6.1.4.1   Rigid Top Protections

Rigid top protections are to have the following characteristics:

- Designed for a uniformly distributed load of 3 kN distributed over an area of 600 × 300 mm (24 × 12 in.)
- Non-slip surface; and
- Opening size not more than 1,500 mm^2 (2 in.^{2}).

Examples of rigid top protections are gratings and plates.

6.1.4.2   Flexible Top Protections

Flexible top protections are to be designed for a central load of 0.03 Rg, but not less than 1 Kn (2,245 lbf) and not more than 3 kN (674 lbf), without making contact with internal fittings or equipment. Examples of flexible top protections are nets/mesh, webbing, and tarpaulin.

Note: *Nets and webbing are to have an opening size not more than (50 × 50) mm (2 × 2 in.).*

6.1.5 Intermediate Cargo Decks

When intermediate cargo decks are fitted, they are to be designed for the following load, applied uniformly distributed on the deck:

$$P_i = \alpha P g \psi$$

where:

α = minimum of 0.5

P = payload

g = acceleration due to gravity

ψ = dynamic factor (= 3).

Intermediate cargo decks are to be designed for a minimum load of half of the total payload (minimum α of 0.5). Values of α other than 0.5 may be modified accordingly.

6.1.6 Design Temperature (T_D)

The design temperature (T_D) is not to be more than -20 °C (-4 °F). When a dedicated service is specified with a lowest daily mean temperature greater than 0 °C (32 °F), the design temperature may be taken as no more than 0 °C (32 °F). This requirement on design temperature to be 0 °C (32 °F) is applicable to regions of temperate climates between 36° North and 36° South and in Australian waters only. Marking requirements for containers operated in these regions is specified in Chap. 9.

6.1.7 Stacking and Stacking Fittings

Offshore containers are not to be stacked during transportation on ships. If an offshore freight or service container is intended for stacking on fixed or floating platforms, the container is to be designed for stacking and stacking height is not to exceed two levels. Empty waste skips with trapezium shaped sides may be designed for stacking at multiple levels. The stacking guides on the top rails or on the underside of a container are to be designed to prevent lateral movement under normal operating conditions.

6.2 Structural Strength

The structural strength of the containers is to be verified through design calculations and tests as indicated in Chap. 2, Sect. 6.4.

6.2.1   Lifting with Lifting Set

6.2.1.1   Loads

The primary structure is to be analysed for a total load of $2.5R_g$, considering an internal load of $(2.5R - T)_g$ evenly distributed over the container floor, where:

R = rating (i.e., the maximum gross mass of the container including permanent equipment and its cargo, but excluding the lifting set)

T = tare mass (i.e., the mass of an empty container including any permanent equipment but excluding cargo and lifting set)

g = acceleration due to gravity.

For tank containers, the actual distribution of the tare mass is to be used.

6.2.1.2   Criteria

Von Mises equivalent stress is not to exceed the allowable stress for plate elements, se, obtained from the following equation:

$$\sigma_e = 0.85C$$

where:

$C = \sigma_y$ for steel

$= \varepsilon$ for aluminium base material

$= 0.7B\sigma_u$ for aluminium heat affected zone

σ_y = specified minimum yield stress

$\beta = 0.8$ for ISO AIMg4.5Mn-HAR/AA5083-H32

$= 0.7$ for all other aluminium alloys and tempers

σ_u = specified minimum tensile strength at room temperature.

Beam elements are to meet the requirements of a standard recognised by Class such as AISC.

6.2.2   Lifting with Forklift Truck

6.2.2.1   Loads

The primary structure is to be analysed for a total load of $1.6(R + S)_g$, considering an internal load of $[1.6(R + S) - T)]_g$ evenly distributed over the container floor. Design force on the primary structure is to be calculated as

$$1.6(R + S)_g.$$

where:

R, T and g are specified in this chapter, Sect. 2.1.1.

Where fork pockets are intended only for handling of the empty container, the design load is to be taken as $1.6(T + S)g$. The weight of the lifting set is to be taken into account when the strength of the fork pockets are to be calculated.

S = mass of the lifting set.

6.2.2.2 Criteria
Von Mises equivalent stress is not to exceed the allowable stress for plate elements, σ_e, indicated in this chapter, Sect. 2.1.2 above. Beam elements are to meet the requirements of a standard recognised by Class such as AISC.

6.2.3 Impact Loads

Impact loads are dynamic loads of very short duration. The ability of a container unit to withstand impact loads is to be determined by calculation. Testing is to be carried out to demonstrate that the container can withstand impact loads (refer to Chap. 7). When simplified calculations are used, each beam is to be considered separately and assumptions concerning support conditions are to be stated in the calculations.

6.2.3.1 Horizontal Impact

6.2.3.1(a) Loads
The primary structure is to be designed to withstand a localised horizontal impact force acting at any point. This force may act in any horizontal direction on the corner post. On all other frame members in the sides, the load may be considered as acting orthogonal to the side. The calculated (static equivalent) stress due to impact is to be combined with the lifting stresses resulting from static lifting forces (Rg). The following loads are to be considered for the calculation of the stress due to impact:

- For corner posts and side rails of the bottom structure: $-0.25Rg$
- For other frame members of the side structure, including the top rails: $-0.15Rg$; and
- For horizontal impact on the tank containers for dangerous cargoes, refer to Chap. 14.

R and g are defined in this chapter, Sect. 2.1.1.

6.2.3.1(b) Criteria
Maximum calculated deflections at these loadings are not to exceed:

(1) For corner posts and bottom side rails: $\ell_n/250$ where is the total length of the rail or post; and
(2) For other frame members: $\ell_n/250$ where ℓ_n is the shortest edge of the wall being considered, in mm (in.).

Note: ℓ_n *is a (nominal) reference length and it is often different from actual span of a beam.*
    For loads indicated in this chapter, Sect. 2.3.1(a), the von Mises equivalent stress is not to exceed the allowable stress for plate elements σ_e, obtained from the following equation:

$$\sigma_e = C$$

where C is defined in this chapter, Sect. 2.1.2. Beam elements are to meet the requirements of a standard recognised by Class such as AISC.

6.2.3.2 Vertical Impact
Maximum vertical impact forces are likely to occur when a container is lowered onto the deck of a heaving supply vessel. If the deck is at an angle, the first impact is likely on a corner. Such impact forces cannot be readily simulated by static forces. Therefore, as an alternative, the strength is to be verified through vertical impact test as described in with Chap. 7, Sect. 6.4. In addition, the following applies.

6.2.3.2(a) Loads
The side rails and end rails in the base are to be able to withstand vertical point forces of $0.25Rg$ at the centre span. R and g are defined in this chapter, Sect. 2.1.1.

6.2.3.2(b) Criteria
Calculated deflections are not to exceed $(\ell_n/250)$, where ℓn is the total length of the rail. Von Mises equivalent stress is not to exceed the allowable stress σ_e, defined in this chapter, Sect. 2.3.1(b). Beam elements are to meet the requirements of a standard recognised by Class such as AISC.

6.2.4 Pad Eye Design

6.2.4.1 Loads
Pad eyes are to be designed to withstand a total vertical force of $3Rg$. The force is to be considered as distributed uniformly between $(n - 1)$ pad eyes, where the value of 'n' is

not to exceed four (4) and not to be less than two (2). The resulting sling force on each pad eye is calculated as follows:

$$F_S = \frac{3Rg}{(n-1)\cos\beta}$$

where:

β = mass of the lifting set.

R and g are defined in this chapter, Sect. 2.1.1.

Containers with only one pad eye may be approved by Class with special consideration; that single pad eye of the container is to be designed for a total vertical force of $5Rg$.

Note: *Containers without a roof may have insufficient strength and stiffness to pass the 2 point lifting test, refer to Chap. 7, Sect. 7.3.3. Therefore, the open top container is to be analysed to withstand the load occurring in the 2-point lifting test. In these calculations, the nominal yield stress of the material should not be exceeded. These calculations do not replace the prototype testing.*

6.2.4.2 Criteria
The maximum concentrated stresses at the hole edge are not to exceed $2\sigma_y$ at design load.

6.2.5 Internal Forces on the Container Walls

The walls of the containers are to withstand a force of $0.6Pg$ (60% of the payload) evenly distributed over the whole surface without experiencing any permanent deformation.

6.3 Other Structural Requirements and Construction

6.3.1 Primary Structure

6.3.1.1 Minimum Material Thickness for Primary Structure
The following minimum material thickness (tmin) requirements apply:

(1) For external parts of the corner posts and bottom rails (i.e., parts forming the outside of the container):
 - For $R \geq 1,000$ kg (2,205 lb) $t_{min} = 6$ mm (0.24 in.)
 - For $R < 1,000$ kg (2,205 lb) $t_{min} = 4$ mm (0.16 in.)
(2) For all other parts of the primary structure: $t_{min} = 4$ mm (0.16 in.); and
(3) Waste skips which are designed to utilise the external skin to support most or all of the load:

- t_{min} = 6 mm (0.24 in.), the area up to 100 mm (4.0 in.) from the side edges.
- t_{min} = 4 mm (0.16 in.), for the remaining part of the side structure.

where:

R = rating (i.e., the maximum gross mass of the container including permanent equipment and its cargo, but excluding the lifting set)

t_{min} = minimum material thickness, in mm (in.).

Note: *The thickness may have to be increased beyond these values to take account of special considerations such as rating, design, corrosion allowances, the need for impact tests of the material,* etc.

6.3.1.2 Pad Eyes

The pad eyes are to align with the sling to the centre of lift, with a maximum manufacturing tolerance of $\pm 2.5°$. The difference in the diagonal measurements between lifting point centres not to exceed 0.2% of the length of the diagonal, or 5 mm (0.2 in.), whichever is the greater. Clearance between the shackle pin and pad eye hole is not to exceed 6% of the nominal shackle pin diameter. The tolerance between pad eye thickness and inside width of shackle are not exceed 25% of the inside width of the shackle. Pad eyes are to be designed as to permit free movement of the shackle and sling termination without fouling the pad eye. Lifting points are to be positioned to minimise the risk of slings fouling against the container or its cargo during operation. If the lifting force is transferred through the thickness of a plate, plates with specified through thickness properties in accordance with a Class recognised standard are to be used.

6.3.1.3 ISO Corner Fittings

Refer to Chap. 15 for requirements.

6.3.2   Secondary Structure

6.3.2.1 Minimum Material Thickness

For Secondary structures made from metallic materials, the required minimum material thickness is:

$$t_{min} = 2 \, mm \ (0.08 \, in.)$$

6.3.2.2 Floors

In the case that there is a possibility that water may collect within the container, provisions are to be provided for drainage.

6.3.2.3   Doors and Hatches

Doors and hatches, including hinges and securing devices, are to be designed to equivalent environmental forces as surrounding structure. Doors are to be capable of being secured by a locking device in the closed position against the structural framing of the opening. Each of the double doors are to be secured by a locking device against the top and bottom structural framing of the opening. All doors are to be capable to be secured in the open position. Locking devices are to be secured in such a manner that they do not allow any disengagement during operations.

6.3.2.4   Internal Securing Points

Containers for general cargo are to have internal securing points that are designed to withstand a force of at least 10 kN (1 tf, 2,248 lbf). A secondary securing device is to be provided for all removal parts.

6.3.3   Additional Structure

6.3.3.1   Forklift Pockets

Forklift pockets may be provided for handling containers in the loaded or unloaded condition. The forklift pockets are to pass completely through the base structure. Forklift pockets are to have closed tops and be provided with means to prevent the container from toppling from the forks. Forklift pockets are to meet the following dimensional requirements:

- The minimum internal dimensions of the forklift pockets are to be 200 mm × 90 mm (8 in. × 3.5 in.); and
- Forklift Pockets are be located as far apart as practicable but need not be more than 2,050 mm (81 in.) or less than 900 mm (35 in.) apart from centre to centre of pockets.

Notes:

(1) *Special requirements apply for fork pockets on tank containers and MEGCs used offshore for the carriage dangerous cargoes, (refer to Chap. 14)*
(2) *The bottom face of the pocket may be fully closed but it is recommended that openings be provided to facilitate maintenance and to minimise the risk of loose items being retained in the pockets which could subsequently fall out during lifting operations. These openings should be dimensioned and positioned so as to minimise the likelihood of the fork tines penetrating or seizing in the opening, or of damaging the free edges at the cut-out*
(3) *In order to compensate for strength reduction in the bottom side rails in the way of forklift pockets, additional strengthening can be placed on the top of the side girders,*

and it is to be in line with the webs of the bottom girder. It is also required to have full penetration welds on the strengthening members and extend at least 100 mm (4 in.) outside the pocket opening at each end. These strengthening members also serve as protections against the forklift truck rupturing the side girder.

6.3.3.2 Tugger Points

Tugger points are the attachments to the container that facilitate handling the container without lifting. They are to be attached to the primary structure of the container only and are to be positioned as low on the structure as possible within the outer edges of the container. Tugger points are to be designed to withstand a load equal to the rating of the container.

6.4   Equipment

Supporting structure of equipment on offshore containers is to be designed for the following factors:

- Dynamic factor: $\psi = 3.0$; and
- Safety factor against breaking: $SF = 2.0$.

External connections of the equipment are to be protected from damage.

Note: *Certification of equipment is not covered in this guide.*

6.4.1   Coating and Corrosion Protection

Coatings, corrosion protection, paint protection and materials of offshore containers are to be suitable for environmental conditions. All offshore container roofs are to be coated with a permanent non-slip medium.

Prototype Testing

7

7.1 Test Equipment and Calibration

Calibrated weights are to be used for verification of the test mass. Calibration of load cell and handset are to be carried out annually in accordance with ISO 7500-1 or similar national or international standards. In all cases, the calibration of test blocks must be performed every 24 months to a national or international standard. Measured mass is to be marked on each block. Calibration certificates for each piece of equipment used in the prototype test of the container must be collected during attendance and retained as part of the prototype test report.

7.2 Prototype Testing

A container selected for prototype testing is to be representative of the manufacturing process to be used in the production of all units. This means it must be constructed in conformity with the approved drawings intended for subsequent production units. Modifications to the offshore container design require engineering revaluation as per Chap. 12 and may require additional prototype testing. Test loads are to be uniformly distributed on the container floor; alternatively, test loads are to be distributed in such a way that it represents the actual load distribution in typical operating conditions. For offshore containers with additional cargo decks, test loads are to be divided between the floor and the additional deck in accordance with Chap. 6, Sect. 6.1.5. In the case that the additional deck is detachable, the test loads are to be placed directly on the floor.

© The Author(s), under exclusive license to Springer Nature Switzerland AG 2025
A. A. Olsen, *Guidance on the Classification of Offshore Containers*, Synthesis Lectures on Ocean Systems Engineering, https://doi.org/10.1007/978-3-031-74857-8_7

7.3   Lifting Test

7.3.1   General

Lifting tests are to be carried out with the lifting set placed at an angle to the vertical equal to the design angle and held for five (5) minutes. Lifting sets used for the prototype testing of the container are not to be used in service. Lifting tests are to be carried out carefully at a gradual speed in order to avoid significant acceleration.

7.3.2   All-Point Lifting

The container under test is to be loaded to a total mass of $2.5R$ and lifted clear off the ground. The test is to be carried out using all the pad eyes and with an internal load that is equal to $2.5R - T$.

7.3.2.1   Acceptance Criteria

• *During Test:* No deflections are to exceed 1/300 of the span of the member; and
• *After Test:* No permanent deformations or other damages are to be observed.

7.3.3   Two-Point Lifting

The container under test is to be loaded to a total mass of $1.5R$ and lifted clear off the ground. The test is to be carried out using only two diagonally opposite pad eyes and with an internal load equal to $1.5R - T$.

7.3.3.1   Acceptance Criteria

• After Test: No permanent deformation or other damage is to be observed after testing.

7.3.4   Post-lifting Test Inspection and Examination of Pad Eye

Non-destructive examination and visual inspection, as per Chap. 8, Sect. 8.2.1, of the pad eyes are to be carried out after the lifting test.

7.4   Vertical Impact Test

7.4.1   General

The containers are to be loaded to their maximum permissible payload (P) and are to be dropped or lowered onto a solid floor. The floor may be covered by wooden planks not exceeding 50 mm (2 in.) in thickness. The container is to be inclined to form an angle with the floor not less than 5 degrees but in no case is the greatest vertical distance between the highest point and lowest point of the underside of the container corners to be more than 400 mm (16 in.). The corner with lowest rigidity is to be the impacted corner. The corner with lowest rigidity is normally at the door end for dry cargo container. One of the following test options is required:

- *Option 1—Drop Test:* For the drop test, the container is to be suspended from a quick release hook and dropped freely for at least 50 mm (2 in.) for an initial impact of at least one (1) m/s; and
- *Option 2—Lowering Test:* For the lowering test, the container is to be lowered to the floor with a minimum constant speed of 1.5 m/s.

Whenever the container is lowered from a crane, the impact speed is to be observed as there is a possibility for the suspending wire and the hook to dampen the free-fall drop speed. In some instances, Class may consider alternative procedures for loading on testing.

7.4.1.1   Acceptance Criteria

- *After Test:* No major deformation or other damage is to be observed.

7.5   Other Tests

7.5.1   Forklift Pockets

Open top containers with an overall length of 6.5 m (21 ft) or more designated to be lifted by forklift pockets while loaded, are to be tested with a total uniform distributed gross mass of 1. 6 $(R + S)g$ and lifted clear of the ground using the fork pockets.

Note: *Forklift pockets testing is not mandatory for other types of containers that have forklift pockets.*

7.5.1.1 Acceptance Criteria

- *During Test:* No deflections are to exceed 1/300 of the span of the member; and
- *After Test:* No significant deformations or other damages are to be observed after the test is carried out.

7.5.2 Driving Ramps

Driving ramps, when fitted to offshore containers, are to be tested for an axle load of 1.25P but need not to be more than 7,260 kg (16,006 lb), evenly distributed between two tyres of a test vehicle. Each tyre is to have a surface area not exceeding 142 cm^2 (22 in^2) with a nominal centre distance of 760 mm (30 in). If the container is specially designed to transport one or more-unit cargoes with a weight (UC) that would give a test axle load higher than 7,260 kg (16,006 lb), the test load is to be two times the UC. For example, if the unit cargo weighs 6,000 kg (13,228 lb), then the test load is to be 6,000 kg (13,228 lb) multiplied twice (i.e., 12,000 kg). Driving ramps are to be clearly marked with the maximum allowable axle load, which shall be 0.8 times the test load.

7.5.3 Stability from Overturning

Refer to section Chap. 6, Sect. 6.1.2.

7.6   Dangerous Goods Cargo

Containers intended for the transport of dangerous goods are to be tested in accordance with the IMDG Code and other relevant Class Rules, international regulations and standards as per Chap. 14.

Production and Manufacture of Offshore Containers

8.1   General

Offshore containers are to be manufactured in compliance with the Class approval letter, approved drawings and specifications, which are to be made available to the Class Surveyor upon their request. The Class Surveyor is to verify that the safety procedures are in place at the manufacturing facility. The Surveyor has the right to decline work if the conditions are considered unsafe in accordance with Class health and safety policy. Offshore containers are to be produced under Class surveillance in a manufacturing facility which has a valid Class Factory Approval Certificate. Offshore containers produced in a facility which does not have a Class Factory Approval Certificate requires 100% inspection. The equipment used to manufacture offshore containers is to be calibrated in accordance with a Class recognised standard. Calibration certificates are to be periodically reviewed by the Surveyor to verify the equipment continues to comply with the applicable standard(s). *Note: Class quality procedures may require calibration certificates for each piece of equipment used in the production of the container to be collected during the annual quality assessment and retained as part of the endorsement.* Welding Procedure Specifications (WPS) and Welder Qualifications (WQ) may be reviewed and accepted by the Class Surveyor prior to production. The Surveyor is to confirm that all WPS' are applicable to the design under manufacture and as approved by Class.

A. A. Olsen, *Guidance on the Classification of Offshore Containers*, Synthesis Lectures on Ocean Systems Engineering, https://doi.org/10.1007/978-3-031-74857-8_8

8.2 Primary Structure

8.2.1 Examination of Welds

8.2.1.1 General

Table 8.1 illustrates the required extent of visual examination of welds. The percentages as specified in the table are applicable for the entire length of the weld. Whenever fuel gas welding is carried out on the primary structure, ultrasonic and magnetic particle examinations are to be carried out along with radiographic examination.

8.2.1.2 Non-destructive Examination (NDE) Methods

The NDE methods specified in Table 8.2 are to be appropriate for the structural welding being inspected.

Table 8.1 Non-destructive examination (NDE) of structural welds

Category of member	Type of examination			
	I Visual examination	II Magnetic particle examination[a]	III Ultrasonic examination[a]	IV Radiographic examination[a]
Essential/ Non-redundant primary structure	100%	100%	100% pad eyes; 20% all other	10%
Non-essential primary structure	100%	20%	20%	10%
Secondary structure	100%	N/A		

Notes
[a] Dye Penetrant examination to be used where magnetic particle examination is not possible

Table 8.2 Standards relevant (NDE) methods

Visual	Magnetic particle	Dye penetrant	Ultrasonic	Radiography
ISO 17637	ISO 17638	ISO 3452–1	ISO 17640	ISO 17636–1; and ISO 17636–2

Note Class B Improved radiographic techniques are to be used. Requests for inspection to alternative NDE methods may be considered by Class on a case-by-case basis

Table 8.3 NDE Acceptance criteria

Visual	Magnetic particle	Dye penetrant	Ultrasonic	Radiography
ISO 5817[a]	ISO 23278	ISO 23277	ISO 11666	ISO 10675–1[b]
Level B	Level 1	Level 1	Level 2	Level 1

Notes
[a] For aluminium ISO 10042
[b] For aluminium ISO 10675-2
[c] Manufacturing facilities may use an alternative standard recognised by class

8.2.1.3 Weld Acceptance Criteria

See Table 8.3.

8.3 Secondary Structure

The Class Surveyor is to verify that the fabrication of secondary structure satisfies the requirements of this Guide and other defined requirements specific to the type of container. The secondary structure with respect to the intended function (cargo securing, prevention of water ingress, etc.) is to be inspected.

8.4 Production Testing

8.4.1 Lifting Test

An all-point test as described in Chap. 7, Sect. 7.3.2 is to be carried out on containers that are randomly selected from the production batch. The number of containers to be tested are to be in accordance with Table 8.4.

Table 8.4 Number of containers required for lifting test

Total number in series	Number to be tested
1–5	1
6–10	2
11–20	3
21–40	4
≥ 40	10% of total

Note The quantity given includes the container which was prototype tested excluding the test sling used for prototype testing

8.4.2   Weather Proofness Testing

If the offshore container is specified to be weatherproof, then weather proofness tests are to be carried in accordance with ISO 1496–1 for at least 10% of the containers in a production series.

Offshore Container Marking and Data Plates

9

9.1 Safety Marking

See Table 9.1.

9.2 Identification Markings

Each container is to be permanently marked with the manufacturer's serial number with characters not less than 50 mm high.

9.2.1 Container Number

In addition to the manufacturer's serial number marking, the container number is to be displayed in contrasting colours as specified below in Table 9.2.

For open sided containers, the container number is to be marked in attached panels specifically to carry this number.

9.3 Information Marking

Each container is to be clearly marked with:

(1) Maximum gross mass (kg)
(2) Tare mass (kg)
(3) Payload (kg)

Table 9.1 Safety marking

	Closed container	Open and framed container	Container with forklift pockets (for empty handling)	Aluminium containers
Location	All around roof perimeter	Top surface of top rail	Near to each fork pocket	All four sides
Text	N/A	N/A	"Empty lift only"	"ALUMINIUM CONTAINER"
Character size (not less than)	Band 100 mm (4 in.) wide	N/A	50 mm (2 in.) high	75 mm (3 in.) high
Colour	Solid contrasting colour	Solid light colour or hatching in a contrasting colour	N/A	N/A

Note In case the roof of the container is recessed below the top rail, the top surface of the top rail is to be marked at a minimum

Table 9.2 Container number

Location	Character size
All sides	75 mm (3 in.) high
Roof (if applicable)	300 mm (12 in.) high or as large as possible if restrictions with availability of space on the roof applies

(4) Relevant electrical hazard classification and zone marking in accordance with Class recognised standards indicated in Chap. 13, Sect. 13.11

(5) Relevant dangerous goods placarding in accordance with the IMDG Code (refer to Chap. 14 for further information); and

(6) The text "ONLY TO BE USED IN TEMPERATE CLIMATES" displayed on the same side of the data plate, whenever the containers are operated in temperate climates as specified in Chap. 6, Sect. 6.1.6 (a), (b), and (c) are to be displayed in characters of a contrasting colour not less than 50 mm (2 in.) high and (f) is to be displayed in characters of a contrasting colour not less than 75 mm (3 in).

Notes:

(1) *An offshore Container carrying dangerous goods which is taken out of service is to have all dangerous goods placarding removed.*

(2) *A matte black panel of appropriate size may be provided for the application of temporary information. It is recommended that this panel be located on a door, where fitted. Other information (e.g., destination) may be added if desired.*

9.4   Marking for Containers with an Intermediate Deck

Payload of the deck is to be displayed on the inside of the container, clearly visible, with characters of contrasting colours not less than 50 mm (2 in.) high.

9.5   Container Data Plate

9.5.1   General

The container data plate is to be:

- Made of non-corrosive materials
- Attached externally to the doors, although for containers with no doors the data plate may be attached in a position that it is clear visible place at all times; and
- Marked in English language with characters not less than 4 mm (0.15 in.) high (provision to include additional languages may be considered).

Note: Aluminium rivets are prohibited for the attachment of data plates.

9.5.2   Contents on the Data Plate

The data plate to be headed:

OFFSHORE CONTAINER DATA PLATE

CLASS O.C: YYYY

(*Where* YYYY *indicates the latest publication year of Class OC Guide*).

En 12,079/Iso 10855

The plate is to indicate the following information:

- Manufacturer's name
- Manufacturer's serial number
- Month and year of manufacture
- Maximum gross mass in kilograms excluding lifting set at the design sling angle
- Tare mass in kilograms
- Payload in kilograms and intermediate deck payload (if applicable)

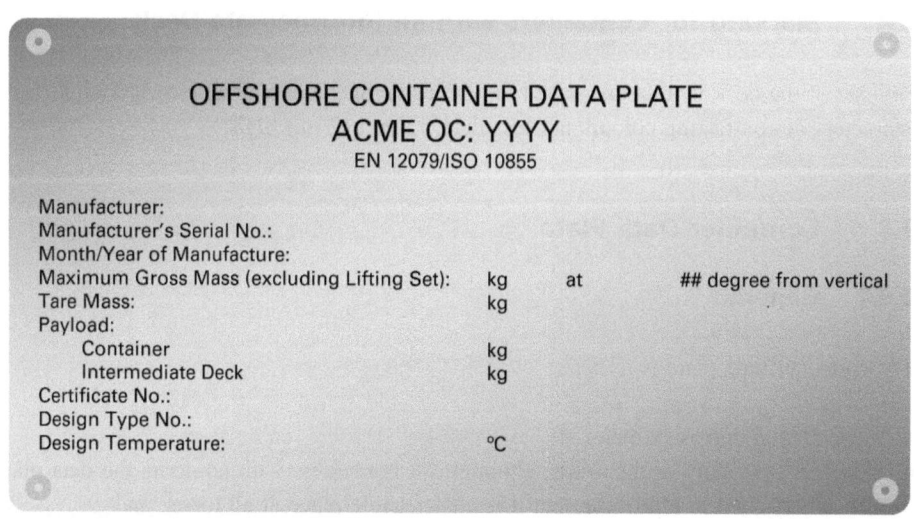

Fig. 9.1 Example of data plate layout

- Production Certificate Number
- Design Type Number; and
- Design temperature.

A sample format of the offshore container data plate described above is shown at Fig. 9.1.

Lifting Sets

10

10.1 General

The lifting set is to be used on a designated container and is to accompany the container at all times. It may be removed only for the purpose of maintenance or replacement. In case there is a necessity for replacement, the new lifting set is to be manufactured to the original specification. It is also necessary to comply with the requirements of a Class recognised standard. The lifting set is to be assembled with components as specified in Chap. 2, Sect. 2.2.1. The quality control procedures of the facility that is intended to manufacture the lifting set is required to comply with requirements specified in Chap. 3.

10.2 Materials and Welding

10.2.1 General

Materials are to be suitable for the intended service conditions. The materials used for manufacturing of the lifting sets are required to comply with the requirements of a Class recognised standard and able to resist dynamic loads.

10.2.2 Identification of Materials

The manufacturer is to adopt a system for the identification of finished plates, shapes, castings, and forgings which will enable the material to be traced to its original heat; and the Surveyor is to be given sufficient documentation and means for verifying the traceability of the material.

A. A. Olsen, *Guidance on the Classification of Offshore Containers*, Synthesis Lectures on Ocean Systems Engineering, https://doi.org/10.1007/978-3-031-74857-8_10

10.3   Materials Used in Slings and Their Components

Materials used in wire ropes, chain slings, ferrules and thimbles are to be in accordance with a Class recognised standard.

10.4   Toughness Requirements

Tests are required to demonstrate that steels meet the Charpy V-Notch (CVN) impact requirements in accordance with ISO 148–1 or a Class recognised standard. The CVN requirements are to be attained when tested at the design air temperature (TD) with minimum average impact energy of 42 J (4.3 kgf-m, 31 ft-lb). For welded components (chains, links etc.) the test is to be sufficient to take impact test samples in the weld with the notch centred in the fusion line with minimum average impact energy of 27 J (2.8 kgf-m, 20 ft-lb). The position of the weld is to be accurately identified by etching with a suitable reagent before cutting the notches. When standard specimens cannot be made, the required energy values are to be reduced as follows:

- 10 mm (0.39 in.) $\times$ 7.5 mm (0.29 in.): 5/6 of the above value; and
- 10 mm (0.39 in.) $\times$ 5.0 mm (0.19 in.): 2/3 of the above value.

10.5   Galvanising

Galvanising of a component is to be carried out under the control of the manufacturer. Materials for galvanised structures are to be fabricated and designed in accordance with industry-recommended practices.

10.6   Material Certificates

The materials used in all components are to be supplied with an inspection certificate in accordance with ISO 10474 or similar Class recognised standard.

10.7   Welding

Before undertaking the welding of any structure subject to the requirements of this guide, a manufacturer is to prove to the satisfaction of the Surveyor that the welding consumables and the proposed process are acceptable by Class and that welders and welding operators are duly qualified for the work intended.

Table 10.1 Required minimum shackle working load limit (WLLs)

Required minimum shackle working load limit (WLLs)		
4 leg sling	2 leg sling	Single leg sling
$WLL_{min}/(3 \times \cos\beta)$	$WLL_{min}/(2 \times \cos\beta)$	WLL_{min}

10.8 Technical Requirements

10.8.1 General Requirements

Slings are to be rated for a maximum angle β of 45 degrees. Hinge type couplings are not typically permitted. The distance between the top master link and the ground when the lifting set is hanging from the top rail is to be not more than 1.3 m (4.3 ft).

10.8.2 Dimensions and Strength of Lifting Sets

The minimum working load limit (WLL_{min}) indicated in Chap. 12, includes dynamic factor which represents the dynamic amplification while lifting operations are carried out in adverse weather conditions.

Note: *For containers in the intermediate container ratings, with values greater than 2,000 kg the WLL_{min} is calculated by interpolation ($WLL_{min} = R \times dynamic\ factor$).*

Selection of lifting set components such as chain, wire rope, shackles and master links are to be in accordance with a Class recognised standard based on the calculated WLL_{min} of the container. Minimum working load limit for each shackle and for each sling (WLLs) are to be determined in accordance with Table 10.1.

Where β is the angle of the sling leg from the vertical and WLL_{min} is the minimum WLL determined from Chap. 12.

10.9 Components

10.9.1 Wire Rope Slings

Wire rope slings are to be constructed to a standard recognised by Class within the restrictions outlined below:

(1) Wire rope is to be 6-stranded and of type 6×19 or 6×36

Note: *In wire rope with 6 × 19 construction, the first number 6 indicates the number of strands, and second number 19 indicate the number of wires to make up one strand. Similarly, for wire ropes with 6 × 36 construction, indicate 6 strands and 36 wires to make a strand* (Fig. 10.1).

(2) The termination of wire rope is to be ferrule secured thimble

(3) Wire ropes are to be either fibre cored or steel cored; and

(4) Wire rope grade 1770/Improved Plowed Steel or 1960/Extra Improved Plowed Steel are to be used.

The working load limit is to be calculated on the basis of the actual rope grade used (Fig. 10.2).

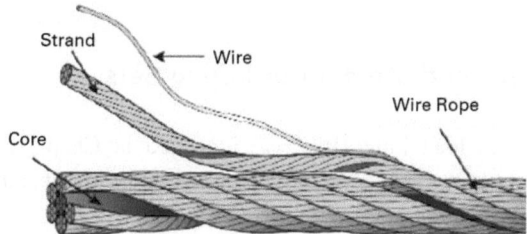

Fig. 10.1 Construction of wire rope

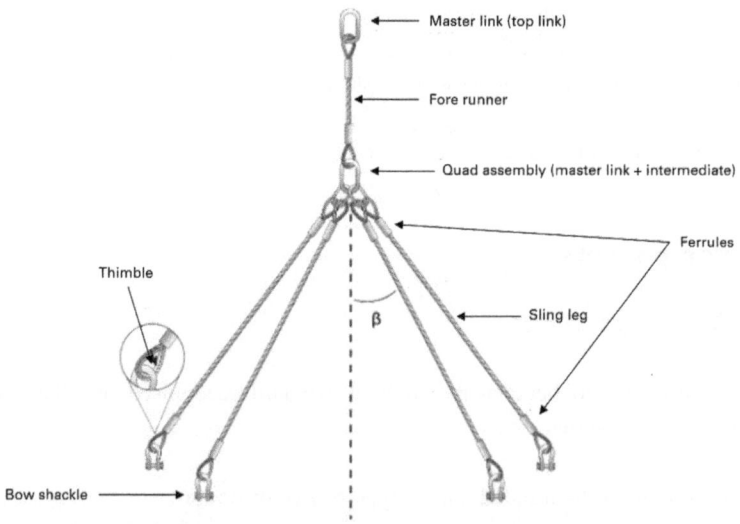

Fig. 10.2 Four-legged wire sling with fore runner

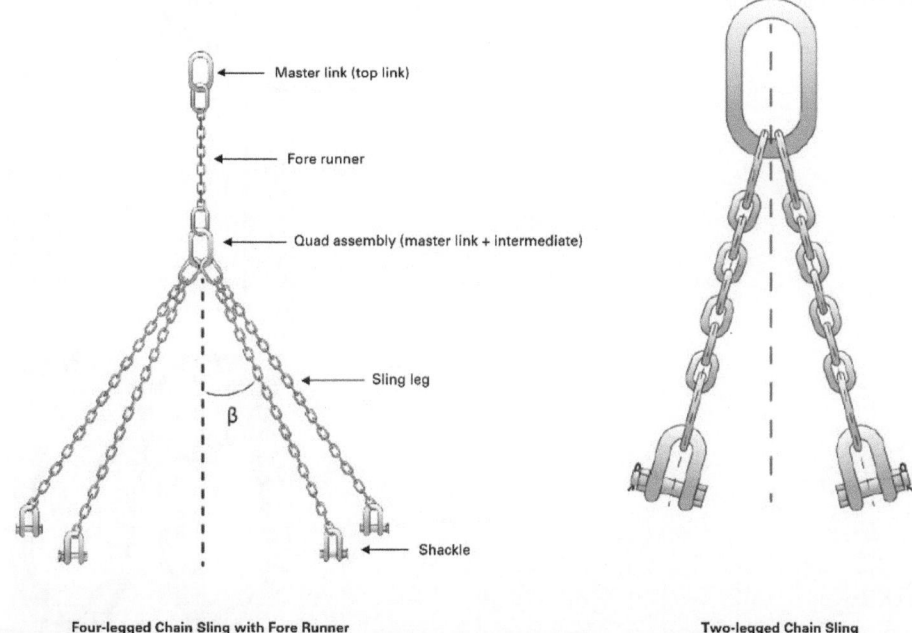

Fig. 10.3 Chain slings

10.9.2 Chain Slings

Chain slings are to be of Grade 8 and are to meet the requirements of the applicable Class recognised standards. Other grades may be accepted after special consideration by Class (Fig. 10.3).

10.10   Master (Top) Link

The master links are to meet the requirements of the applicable Class recognised standard (Fig. 10.4).

10.10.1 Ferrules

Ferrules are to be manufactured to a standard recognised by Class.

Fig. 10.4 Master link
assembly

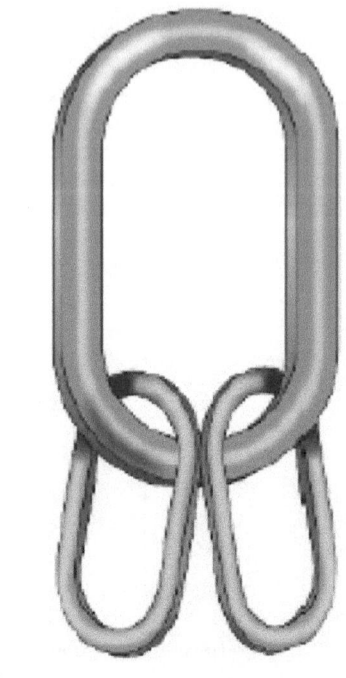

Fig. 10.5 Thimble

10.10.2 Thimbles

Thimbles are to be manufactured to a standard recognised by Class (Fig. 10.5).

10.10.3 Shackle

Shackles are to be manufactured to a standard recognised by Class and with the additional requirement that the tolerance on the nominal diameter of the shackle pin to be –0/ + 3%. Shackles are restricted to bolt type pin with hexagon head, hexagon nut, and split cotter pin only (Fig. 10.6).

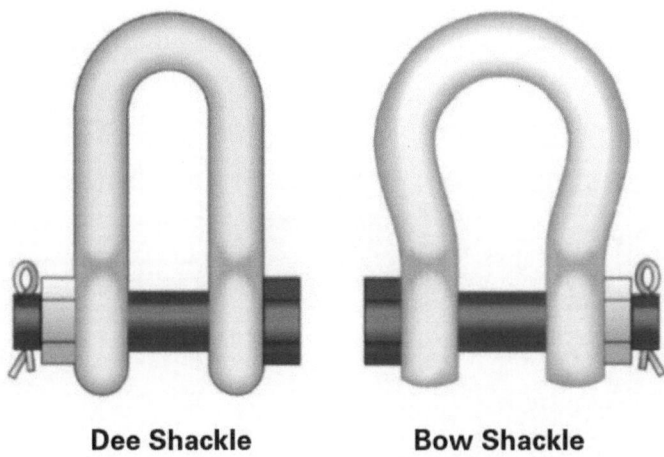

Dee Shackle   **Bow Shackle**

Fig. 10.6 Shackles

10.11 Marking

The various components of the lifting sets are to be marked in accordance with standards used in the design approval of the component.

10.11.1 Shackles

Shackles are to be permanently marked with a unique identification whenever they are fitted to a sling as a permanent assembly. The marking is to be positioned in such a way that it is away from high tensile stress areas with characters of at least 5 mm (0.2 in.) high applied using "low stress" stamps.

10.11.2 Slings

A permanently marked identification tag made of metal is to be secured to the top assembly of the sling. For chain slings, the tag is to be 8-sided, similar to Fig. 10.7, and for wire rope slings, the tag is to be round, similar to Fig. 10.8. The tags for the slings are to be marked with the following information:

- Class OC mark
- Unique identification number of the sling
- Number of legs

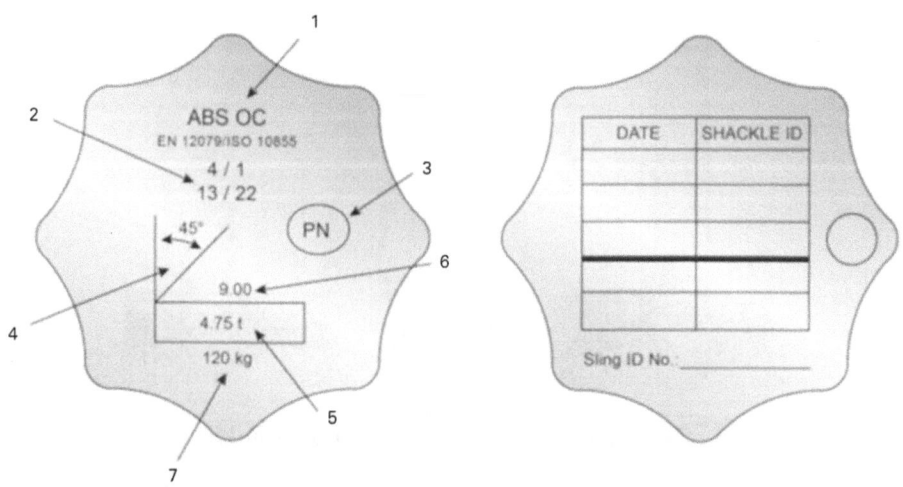

(1)   Class OC
(2)   4 Legs of 13 mm, 1 forerunner of 22 mm (example)
(3)   Manufacturer Mark
(4)   Sling angle
(5)   Shackle size
(6)   WLL (t)
(7)   Mass of the lifting set

Fig. 10.7 Identification tag for chain sling

- Diameter of chain or wire rope used, including the top leg where fitted
- Working load limit (WLL) in tonnes
- Maximum angle of the sling legs from the vertical; and
- Mass of the lifting set (S).

10.11.2.1 Marking Alternatives

As an alternative to marking slings with a tag, one of the following methods may be used for marking:

(1) The marking required by Chap. 10, Sect. 10.10 may be marked on a ferrule on wire rope slings; and
(2) Slings may be marked with a small tag with only an ID number on it. All other information required by Chap. 10, Sect. 10.10 shall be available, either electronically or by other means.

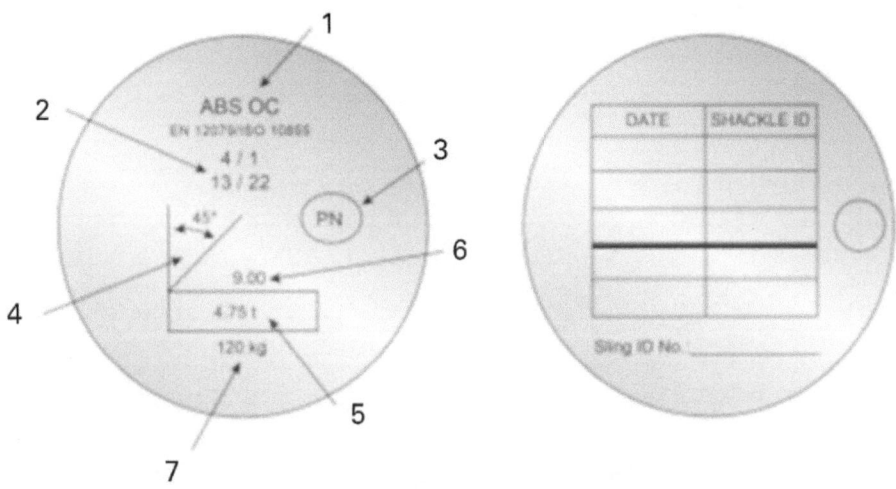

(1) Class OC
(2) 4 Legs of 13 mm, 1 forerunner of 22 mm (example)
(3) Manufacturer Mark
(4) Sling angle
(5) Shackle size
(6) WLL (t)
(7) Mass of the lifting set

Fig. 10.8 Identification tag for wire rope sling

Note: *Dropped objects such as an identification tag on an offshore container lifting set can be a major health and safety issue. The risk of such accidents can be reduced or removed if one of the alternative solutions are used. However, national authorities or other stakeholders might not accept such alternative marking.*

In-Service Inspections

11

11.1 General

This part of the guide specifies requirements for the periodic inspection, examination and testing of offshore freight, waste skips, and service containers built in accordance with this guide. Inspection requirements following damage and repair of offshore containers are included in Chap. 13. Class requires the container to be periodically inspected, examined and tested in the presence of an Class Surveyor in accordance with the requirements detailed in this chapter, Sects. 11.5 and 11.6. The Class Surveyor has the authority to request additional inspections, examinations and/or tests in order to provide confidence the container or lifting set is satisfactory for service. Compliance with the schedule of periodic inspections, examinations and tests for offshore containers certified to this guide is the responsibility of the owner or operator. The periodic inspection, examinations or tests may be performed within one (1) month of the test date without affecting the schedule of inspection, examination and tests. Any container or lifting set which has exceed the test date is to be removed from service until the applicable tests have been satisfactorily performed to the requirements of this guide. Containers and lifting sets which are out of test for more than six (6) months are to be tested in accordance with the test criteria at the four (4) year interval indicated in Tables 11.1 and 11.2.

11.2 Schedule of Inspection for Offshore Containers

See Table 11.1.

T: indicates load test; non-destructive examination, and visual examination; or;

V: indicates visual inspection only; or;

VN: indicates NDE and visual inspection

© The Author(s), under exclusive license to Springer Nature Switzerland AG 2025   59
A. A. Olsen, *Guidance on the Classification of Offshore Containers*, Synthesis Lectures on Ocean Systems Engineering, https://doi.org/10.1007/978-3-031-74857-8_11

Table 11.1 Schedule of inspection, examination and tests for offshore containers

Time or interval (max. interval)	Inspection, examination and tests			
	Lifting test	NDE	Visual inspection	Suffix to be marked on inspection plate
Initial certification	Comply with all the requirements specified in Chap. 2			
12 months	N/A [b]	N/A [b]	Yes	V
48 months	N/A [b]	Yes	Yes	VN
Substantial repair or alteration	Yes	Yes	Yes	T

Notes
[a] Class is the sole judge in determining which repairs or alterations require testing.
[b] Class may require other or additional inspections, examinations and or tests.

Non-destructive examination and a visual inspection are to be carried out after the load test

11.2.1 Schedule of Periodic Inspection, Examination and Test—Offshore Containers

11.2.2 Lifting Test

Load tests are to be in accordance with Chap. 8, Sect. 3.2 of this guide.

11.2.3 Non-destructive Examinations

Non-destructive examination is to be in accordance with Chap. 9, Sect. 2.1 of this guide.

11.2.4 Visual Inspection

The Class Surveyor is to perform a visual inspection on the interior and exterior of an unloaded container. Means for inspection of all load bearing parts, including the underside of the base structure, are to be provided. Access to areas around fixed equipment or other obstructions are to be provided to the satisfaction of the Class Surveyor. The following are to be visually inspected to verify that they are free from visible defects, excessive corrosion, distortion, mechanical damage or any other signs of distress or overload:

- Flooring
- Lashing points
- Pad eyes
- Structure; and
- Welds.

The above is not a comprehensive list and does not include all possible items. Additionally, the following are to be visually inspected to verify that they are functioning as intended:

- Door Closures
- Drainage facilities
- Hatches
- Locking devices; and
- Other appendages.

The above is not a comprehensive list and does not include all possible items.

11.3   Schedule of Inspection, Examination and Tests—Lifting Sets

See Table 11.2.

T: indicates load test; non-destructive examination, and visual examination; or;

V: indicates visual inspection only; or;

VN: indicates NDE and visual inspection

Non-destructive examination and a visual inspection are to be carried out after the load test.

11.3.1   Load Test of Chain Sling Legs

A test load equal to $2 \times$ WLL of a single leg, $\pm 2\%$, is to be applied to each leg without shock. The load is to be applied for at least 2.5 min before measurements are taken.

Table 11.2 Schedule of periodic inspection, examination and testing of lifting sets

Time or interval (max. interval)	Applicable to:	Inspection, examination and tests			
		Load test	NDE	Visual inspection	Suffix marked on sling tag
Initial certification	Complete lifting set	Comply with all the requirements in this chapter			
12 months	Complete lifting set	N/A	N/A	Yes	V
48 months	Sling components and joining links excluding legs	Load test or NDE		Yes	T or VN*
	Chain sling legs	Load test or NDE		Yes	T or VN*
	Shackles	N/A	N/A	Yes	V
	Wire rope legs	N/A	N/A	Yes	N/A
Substantial repair or alteration	Complete lifting set	Yes	Yes	Yes	T

Note Dependent upon whether tested or examined

11.3.2 Non-destructive Examination of Sling Components Except Wire Rope Legs

Magnetic particle examination is to be undertaken as specified in Chap. 9, Sect. 2.1.

11.3.3 Visual Inspection of the Lifting Set

The Class Surveyor is to perform a visual inspection on the lifting set components in accordance with Class recognised standards to verify that they are free from visible defects, excessive corrosion, distortion, mechanical damage or any other signs of deterioration or overload.

11.4   Inspection Plates/Tags

11.4.1 Offshore Containers Inspection Plate

Containers plates are to be fitted with a plate carrying all the information specified in this chapter. The plate is to be as follows:

- Made of non-corrosive material
- Attached externally to the door in a manner to avoid unauthorised or accidental removal (for containers with no doors the data plate is to be attached in a clear visible place); and
- Marked permanently in English language with characters not less than 4 mm high (additional languages may be considered by Class).

Aluminium rivets have been found to be unsuitable as a fixing method in the offshore environment and are not used.

11.4.2 Contents of the Inspection Plate

The containers certified by Class are to have an Inspection plate that is headed:

OFFSHORE CONTAINER INSPECTION PLATE

CLASS O.C:YYYY

EN 12079 / ISO 10855

The plate is to indicate the following information:

- Owner's container number
- Owner's name; and
- Date of last inspection.

The date of last inspection is to be the date on which the most recent inspection was carried out to the satisfaction of the Class Surveyor. The date of next inspection is not to be indicated on the data plate. Provision to facilitate permanent marking to record a minimum of nine inspections is to be provided.

11.4.3 Marking

Upon satisfactory completion of the inspection, examination and when applicable, test(s), the plate is to be permanently marked, in accordance with Table 11.1, as follows:

- The date (YYYY-MM-DD) of the inspection, examination, and when applicable, test(s) together with the unique identification mark of the competent person together with either:
- Suffix T; indicating a lifting test, non-destructive examination and visual inspection; or

- Suffix VN; indicating non-destructive examination, and visual inspection; or
- Suffix V; indicating visual inspection only.

Notes:

(1) *For marking of the inspection plate. Further information, refer to this chapter, 4.2.*
(2) *A recommended format for the plate is shown in Fig. 11.1.*
(3) *The information required for the inspection plate may be combined with the offshore container data plate.*

Inspection plate and the Data Plate can be combined into a single plating as below in Fig. 11.2.

11.5   Inspection Tag for Lifting Set

11.5.1 Marking

Upon satisfactory completion of all inspections, examinations and tests, the sling identification tag is to be attached by suitable means to prevent accidental removal and permanently marked in accordance with Table 11.2 as follows.

The date YY-MM-DD of the inspection/test, as applicable, together with the unique identification mark of the Class together with either:

- Suffix T: indicating load test; non-destructive examination, and visual examination; or
- Suffix V: indicating visual inspection only; or
- Suffix VN: indicating NDE and visual inspection.

11.6   Existing Offshore Containers

11.6.1 General

Existing containers that have not previously been certified by Class according to the provisions of this guide may be considered for certification. Contact the Class Corporate Container Certification group for complete details. The certification procedure will include general compliance with this guide and as agreed by Class. The certification procedure will include a technical design review, inspection and testing of the existing design prior to issuance of the applicable certification. This certification process may be applied on a case-by-case basis for the purpose of bringing existing containers into the Class in-service inspection program per Chap. 12 of this guide.

Fig. 11.1   Example of inspection plate layout

11.6.2 Marking and Date Plate

Container markings and date plates are to be in accordance with Chap. 9.

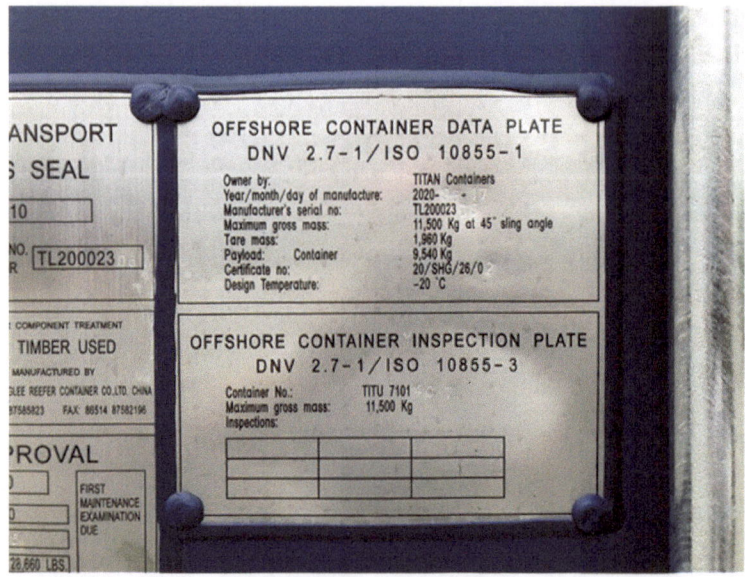

Fig. 11.2 Template of combined data plate and inspection plate

11.7 Emblem

The Class emblem is a representation that will be affixed to each existing offshore container that meets the criteria contained in this document and is approved by Class.

Repair and Modification Procedures

12

12.1 General

Periodic inspections are required by this guide to determine whether damage or wear sustained during operation is extensive enough to require the container or lifting set to be repaired or replaced. Damage and wear is generally measured against the original condition. The Class Surveyor is the sole judge with regards to certification under this guide in determining the extent of the damage or wear and whether the container or lifting set is to be repaired. A container or lifting set which is considered damaged to the extent that a repair is required, is to be removed from service until repairs are carried out and inspected by a Class Surveyor. Means to transport the container or lifting set are to be arranged which does not include the utilisation of the damaged components of the container or lifting set. The repair facility is to operate under an established quality control system in accordance with the requirements of this guide. The Class Surveyor is to review and accept the Welding Procedure Specifications (WPS) and Welder Qualifications being used for the repair. Repairs are to be carried out in accordance with the requirements of this guide. The damaged or worn areas are to be repaired to their original dimensions and profiles in equivalent grades of material. The Class Surveyor is to review all records detailing the repair to verify that the container or lifting set has been repaired to the original condition. The container or lifting set is to be tested and inspected in the presence of a Class Surveyor in accordance with Chap. 11, Tables 11.1 or 11.2 of this guide.

© The Author(s), under exclusive license to Springer Nature Switzerland AG 2025   69
A. A. Olsen, *Guidance on the Classification of Offshore Containers*, Synthesis Lectures on Ocean Systems Engineering, https://doi.org/10.1007/978-3-031-74857-8_12

12.2   Modification Procedures

Modifications or revisions to an existing design are to be submitted to Class in accordance with Chap. 2–10 and/or 11. The facility modifying the container or lifting set is to operate under an established quality control system in accordance with the requirements of this guide. The Class Surveyor is to review and accept the Welding Procedure Specifications (WPS) and Welder Qualifications being used for the repair. The Class Surveyor is to review records detailing the modification to verify if the container or lifting set has been modified in accordance with the approved drawings. The container or lifting set is to be tested, if required by the Class approval letter, and inspected in the presence of a Class Surveyor in accordance with Chaps. 2–10 and/or 11.

12.3   Recordkeeping

The owner is to retain all records of approval and inspection for traceability of the container or lifting set during its life cycle. The records are to include, but are not limited to, the following:

- Engineering approval
- Prototype testing certificate
- Production records
- Production certificate
- Records of repair
- Records of modification
- Changes in identification; and
- Transfer of ownership.

All records supporting the certification of the container or lifting set are to be made available to the Class Surveyor upon their request.

Class Recognised Standards 13

13.1 General

The following list consists of standards which are used in industry and are recognised by Class. Additional standards may be recognised contingent on a review to determine whether the standard is at least equivalent to that recognised by Class.

13.2 Materials

ISO 6892–1 Metallic materials—Tensile testing—Part 1: Method of test at room temperature.

ISO 148–1 Metallic materials—Charpy impact test—Part 1: Test method.

ISO 10474 Steel and steel products- Inspection documents.

EN 10,204 Metallic Products-Types of inspection documents.

EN 10,025–1 Hot-rolled products of structural steels—Part 1: General Technical delivery conditions.

EN 10,025–2 Hot-rolled products of structural steels—Part 2: Technical delivery conditions for non-alloy structural steels.

EN 10,025–3 Hot-rolled products of structural steels—Part 3: Technical delivery conditions for normalised/normalised rolled weldable fine grain structural steels.

EN 10,025–4 Hot-rolled products of structural steels—Part 4: Technical delivery conditions for thermos mechanical rolled weldable fine grain structural steels.

EN 10,210–1 Hot finished structural hollow sections of non-alloy and fine grain structural steels. Technical delivery requirements.

EN 10,219–1 Cold formed welded structural sections of non-alloy and fine grain steels. Technical delivery requirements.

© The Author(s), under exclusive license to Springer Nature Switzerland AG 2025   71
A. A. Olsen, *Guidance on the Classification of Offshore Containers*, Synthesis Lectures on Ocean Systems Engineering, https://doi.org/10.1007/978-3-031-74857-8_13

EN 10,088–2 Stainless Steels—Part 2: Technical delivery conditions for sheet/plate and strip of corrosion resisting steels for general purposes.

ASTM A240/240 M Standard specification for chromium and chromium—nickel stainless steel plate, sheet, and strip for pressure vessels and for general applications.

EN 10,250–2 Open die steel forgings for general engineering purposes—Part 2:Non-alloy quality and special steels.

EN 10,250–3 Open die Steel Forgings for General Engineering purposes—Part 3: Alloy special steels.

ISO 209–1 Aluminium and Aluminium Alloys—Chemical Composition.

13.3   Quality Control for Manufacturing Facilities

ISO 9001 Quality Management Systems—Requirements.

13.4   Iso Corner Fittings

ISO 1161 Series 1 freight containers—Corner fittings—Specification.

ISO 668 Series 1 freight containers—Classification, dimensions, and ratings.

13.5   Padeyes

EN 10,164 Steel products with improved deformation properties perpendicular to the surface of the product—Technical delivery conditions.

13.6   Welding

EN 9606–1 Qualification test of welders. Fusion welding. Steels.

EN ISO 9606–2 Qualification test of welders. Fusion welding-Part 2: Aluminium and Aluminium alloys.

ASME Boiler and Pressure Vessel Code, Section IX.

AWS D1.1 Structural welding code—Steel.

ISO 15607 Specification and qualification of welding procedures for metallic materials—General rules.

ISO 15609–1 Specification and qualification of welding procedures for metallic materials—Welding procedure specification—Part 1: Arc welding.

ISO 15614–1 Specification and qualification of welding procedures for metallic mate-rials—Welding procedure test—Part 1: Arc and gas welding of steels and arc welding of nickel alloys.

ISO 15614–2 Specification and qualification of welding procedures for metallic materials—Welding procedure test—Part 2: Arc welding of Aluminium and its alloys.

EN ISO 15613 Specification and qualification of welding procedures for metallic materials—based on preproduction welding test.

13.7 Calibration

EN ISO 7500–1 Metallic materials. Verification of static uniaxial testing machines. Ten-sion/ compression testing machines. Verification and calibration of the force—measuring system.

13.8 Non-destructive Testing

ISO 17637 Non-destructive testing of welds. Visual testing of fusion –welded joints.

ISO 17638 Non-destructive testing of welds. Magnetic particle testing.

EN ISO 3452–1 Non-destructive testing. Penetrant testing-Part 1: General Principals.

ISO 17640 Non-destructive testing of welds. Ultrasonic testing- techniques, testing levels and assessment.

EN ISO 17636–1 Non-destructive testing of welds. Radiographic testing. X- and gamma- ray technique with film.

EN ISO 17636–2 Non-destructive testing of welds. Radiographic testing. X- and gamma- ray technique with digital detectors.

ISO 23278 Non-destructive testing of welds. Magnetic particle testing of welds-Acceptance levels.

ISO 23277 Non-destructive examination of welds. Penetrant testing of welds. Accep-tance levels.

EN ISO 17643 Non-destructive testing of welds. Eddy current testing of welds by complex—plane analysis.

EN ISO11666 Non-destructive testing of welds. Ultrasonic testing—Acceptance levels.

EN ISO 5817 Welding. Fusion-welded joints in steel, nickel, titanium and their alloys (beam welding excluded).

Quality levels for imperfections.

Chapter 4 In-Service Inspections.

Appendix 1 CLASS Recognised Standards 4-A1

ISO 10675–1 Non-destructive testing of welds. Acceptance levels for radiographic testing. Steel, nickel, titanium and their alloys.

ISO 10042 Arc-welded joints in aluminium and its alloys—Quality levels for imperfections.

ISO 10675–2 Non-destructive testing of welds. Acceptance levels for radiographic testing. aluminium and its alloys.

EN ISO 3834–2 Quality requirements for fusion welding of metallic materials—Part 2: Comprehensive Quality Requirements.

ISO 9712 Non-destructive testing. Qualification and certification of NDT personnel.

13.9   Weatherproofness Testing

ISO 1496–1 Series 1 freight containers—Specification and testing—Part 1: General Cargo containers for general purposes.

13.10   Tank Containers

ISO 1496–3 Series 1 freight containers—Specification and testing. Tank containers for liquids, gases and pressurised dry bulk.

13.11   Containers for Bulk Solids

ISO 1496–3 Series 1 freight containers—Specification and testing. Tank containers for liquids, gases and pressurised dry bulk ISO 1496–4 Series 1 freight container. Specification and testing. Non-pressurised containers for dry bulk.

13.12   Hazardous Areas

ATEX Directive Equipment intended for use in Potentially Explosive Atmospheres IECEx International Electrotechnical Commission for use in Explosive atmospheres.

13.13   Analysis—Beam Elements

Manual of Steel Construction, AISC, 9th Edition.

13.14   Standards for Lifting Set and Their Components

ASME B30.9 Slings—Safety Standard for Cableways, Cranes, Derricks, Hoists, Hooks, Jacks, and Slings.

ASME B30.26 Rigging Hardware—Safety Standard for Cableways, Cranes, Derricks, Hoists, Hooks, Jacks, and Slings.

ISO 898–1 Mechanical properties of fasteners made of carbon steel and alloy steel—Part 1: Bolts, screw and studs with specified property classes—coarse thread and fine pitch thread API Spec. 9A Specification for Wire Rope.

EN 12,385–4 Steel wire ropes. Safety. Stranded ropes for general lifting applications.

EN 13,411–3 Terminations for steel wire ropes. Safety. Ferrules and ferrule—securing.

EN 13,411–1 Terminations for steel wire ropes. Safety. Thimbles for steel wire rope slings.

EN 818–4 Short link chain for lifting purposes—Safety—Part 4—Chain slings —Grade 8.

EN 818–6 Short link chain for lifting purposes. Safety. Chain slings. Specification for information for use and maintenance to be provided by the manufacturer.

EN 1677–1 Components for slings. Safety. Forged steel components, Grade 8.

EN 13,414–1 Steel wire rope slings. Safety. Slings for general lifting service.

EN 13,414–2 Steel wire rope slings. Safety. Specification for information for use and maintenance to be provided by the manufacturer.

ABNT NBR 13,545 Lifting purposes—Shackles (Brazilian Standard).

ISO 2415 Forged shackles for general lifting purposes—Dee shackles and bow shackles.

EN 13,889 Forged steel shackles for general lifting purposes. Dee shackles and bow shackles Grade 6. Safety.

Certification Requirements for Offshore Tank Containers/Portable Tanks and MEGC

14

14.1 Tank Containers for Dangerous Goods

All tank containers intended for marine transport of dangerous goods are to be certified to the International Maritime Dangerous Goods Code (IMDG Code). Tank containers built after 1 January 2003 must be built and certified in accordance with the requirements for UN Type tanks (Chap. 6.7 in the IMDG Code). Tank containers built before that date may be in accordance with Chap. 13 of the older IMDG code (i.e., up to amendment 29 of that code) Tank containers that are certified according to these requirements are also allowed for road and rail transport.

14.1.1 Tank Containers—General

In addition to complying with other relevant design codes and requirements, tank containers are to be suitable for offshore service.

14.1.2 Frame

Tank container frames are to be designed to protect the tank/elements and the equipment.

14.1.3 Tanks for Fluids

Tanks design are to conform to the relevant sections of ISO 1496–3. Tanks for dangerous cargoes are to fulfil the requirements of the IMDG Code and are to be designed according

© The Author(s), under exclusive license to Springer Nature Switzerland AG 2025 77
A. A. Olsen, *Guidance on the Classification of Offshore Containers*, Synthesis Lectures on Ocean Systems Engineering, https://doi.org/10.1007/978-3-031-74857-8_14

to recognised standards such as ISO 1496–3 or to Class recognised standards. A tank and its support are to be able to withstand lifting and impact loads. In addition, due account is to be taken of fluid surge arising from partly filled tanks. *Note*: *The IMDG Code has restrictions for loaded handling of tanks over a certain length, by forklift. Reference should be made to Chaps. 4.2 and 6.7 of the IMDG Code.*

14.1.4 Impact Protection on Tank Containers for Dangerous Cargoes

All parts of the tank containers are to be protected from impact damage and in addition to the requirements of Chap. 7, Sect. 7.2.3 of this guide, the following requirements apply:

(1) Protective covers or housings are to be provided to minimise the risk of damage to the tank containers top and its fittings due to impact
(2) A clearance of at least 100 mm (4 in.) is to be provided between external fittings and the top of the container
(3) Measures are to be taken to prevent any part of the lifting set from fouling the fittings, manholes, cleats or other protrusions on the tank
(4) Protective beams are to be provided to give necessary protection to the tank shell when the same is nearest to the outer plane of the sides
(5) The residual clearance of at least 10 mm (0.4 in.) is to be provided between the member and any part of the tank shell or its fittings at the maximum calculated elastic deflection of any side member; and
(6) The underside of the tank shell (including sumps) and bottom valves or other fittings, are to extend below a level 150 mm (6 in.) above the bottom of the framework (the underside of the side or end rails). Any such part extending below 300 mm (12 in.) above the bottom of the framework, is to be protected by beams or plating.

Special considerations are to be given to tank containers with direct connections between the tank and the framework.

14.1.5 Periodic Inspections for Tank Containers

Offshore tank containers that are certified for dangerous goods according to IMDG code are subject to five-year periodic inspections and tests, and to intermediate 2.5-year periodic inspections and tests.

14.1.6 Multiple Element Gas Containers (MEGCs)

Multiple Element Gas Containers (MEGC) are multimodal assemblies of pressure receptacles or elements that are interconnected by a manifold for filling and discharge which are assembled within an ISO framework and include service equipment necessary for the transport of gases. Additional information on this type of container can be found in the latest edition of Class Rules for the certification of cargo containers.

Container Corner Castings

15

15.1 General

Container corner castings are required to be certified in accordance with the requirements prescribed in this Chapter. The certification procedure is indicated in Chap. 2, Sect. 2.15.2. Corner fittings are to conform to the requirements of ISO 1161. Lifting offshore containers with shackles in corner fittings is not acceptable. Corner fittings of unique design for special purpose containers will also be considered for certification provided the strength requirements are not less than those specified by ISO Standard 1161. Corner castings are to be produced under Class Surveyor witness in a manufacturing facility which has a valid Class Factory Approval Certificate. Corner castings produced in a facility which does not have a Class Factory Approval Certificate will require a certification by heat treatment lot specified in the Class Rules for the certification of cargo containers. Class may accept certification of ISO Corner Fittings by other IACS Societies under special considerations provided that all the data and supporting document are submitted for review. Corner fittings are not allowed to carry out lifting operations on offshore containers.

15.2 Process of Manufacture

The steel is to be made by the open-hearth, electric furnace, or basic oxygen process. Other processes of manufacture will be specially considered.

15.3 Heat Treatment

All castings are to be either fully annealed, normalised, or normalised and tempered.

A. A. Olsen, *Guidance on the Classification of Offshore Containers*, Synthesis Lectures on Ocean Systems Engineering, https://doi.org/10.1007/978-3-031-74857-8_15

Table 15.1 Chemical composition (%)

Carbon (max.)	Manganese (max.)	Silicon (max.)	Sulphur (max.)	Phosphorous (max.)	Aluminium[a] (min.)
0.20	1.20	0.50	0.035	0.035	0.015

Notes
[a]Aluminium may be partly or totally by other fine graining elements as stated in the approved specifications
[b]The manganese may exceed 1.20% provided that the carbon content plus one-sixth of the manganese content does not exceed 0.45%
[c]Residual elements are not to exceed 0.80%
[d]Residual elements individual % maximum (Cu = 0.30, Cr = 0.30, Ni = 0.40, Mo = 0.15)

15.4   Material Specifications

Corner castings are to be made of carbon steel according to the chemical and mechanical properties listed in this chapter, Sects. 4.2 and 5.3 of this guide. Other material specifications submitted for certification of corner castings will be specially considered.

15.4.1 Chemical Analysis

An analysis of each heat of steel is to be made by the manufacturer to determine the percentages of the elements specified below. The chemical analysis is to be made from a sample taken during the pouring of the heat. If drillings are used from a finished casting, they are to be taken not less than 6 mm (0.24 in.) beneath the surface. The chemical composition thus determined is to conform to the requirements prescribed in Table 15.1 of this guide. Chemical analysis certificates are to be provided to the Surveyor.

15.4.2 Chemical Requirements

See Table 15.1.

15.5   Tension Test

One tension test is to be performed on a specimen from each heat treatment lot. The tension test is to be performed in accordance with the American Society for Testing and Materials (ASTM) Standard A 370—Mechanical Testing of Steel Products, or equivalent. The mechanical properties thus determined are to conform to the requirements specified below in the chapter, Sect. 5.3 of this guide.

15.5.1 Tensile Test Specimen

Test bars are to be poured in special blocks, similar to those shown in ASTM A 370, from the same heat as the casting represented, and are to be heat treated in production furnaces to the same procedure as the castings they represent. Alternatively, test coupons may be cut from the heat treated castings or cast integrally. Test specimens are to be machined to the form and dimensions shown in ASTM A 370. If any specimen is machined improperly or if flaws are revealed by machining or during testing, the specimen may be discarded and another substituted from the same heat treatment lot.

15.5.2 Retests

If the results of the mechanical tests do not conform to the requirements specified, heat-treated castings may be reheat-treated and retested, but not more than twice.

15.5.3 Tensile Properties

- Minimum yield strength 240 N/mm^2 (24.5 kgf/mm^2, 34,809 psi)
- Tensile strength $450–600 \text{ N/mm}^2$ (46 to 61.2 kgf/ mm^2, 65,267 to 87,022 psi)
- Minimum elongation in 50 mm (2 in.)   25%
- Minimum reduction in area   40%

15.5.4 Charpy Impact Test

Charpy impact test properties are to be determined on each heat from a set of three Charpy V-notch specimens made from a test coupon in accordance with ASTM A370 and tested at a test temperature of -20 °C (-4 °F). The acceptance requirements are to be the value of energy absorbed. The minimum average Chap. 4. In-Service Inspections absorbed energy value of three specimens is to be 27 Joules (20 ft lbs), and one individual value may be below the average value but shall not be lower than 70% of the average.

15.6   Inspections

15.6.1 Dimensional Inspection

Each casting is to be inspected by the manufacturer to insure compliance with the dimensional requirements defined in ISO 1161. Satisfactory records of such inspection are to be available to the Surveyor.

15.6.2 Visual Inspection

Each casting is to be inspected by the manufacturer for general appearance and surface defects. The castings are to be free from defects. Satisfactory records of such inspections are to be available to the Surveyor.

15.6.3 Internal Discontinuities Examination

One casting from each 400 (50 sets) are to be examined by the manufacturer for internal discontinuities using either radiographic or ultrasonic methods.

15.6.3.1 Radiographic Examination

Castings are to be examined for internal discontinuities by means of X-ray or gamma rays. The procedure is to be in accordance with ASTM Recommended Practice E 94 and Method E 142. The types and degrees of discontinuities considered are to be judged by ASTM Reference Radiograph E 446. Basis for acceptance is to be as follows:

Nature of defects	Radiographic acceptance criteria
Blow holes	Level 4
Inclusions	Level 4
Shrink holes category	CA, AB, CC, or CD level 3
Cracks	None
Quench cracks	None

15.6.3.2 Ultrasonic Inspection

Castings are to be examined for internal discontinuities by means of ultrasonic inspection. The inspection procedure is to be in accordance with ASTM Specification A 609. Methods of testing and basis of acceptance are to be agreed upon.

15.7   Marking

Each corner casting will be identified with the foundry identification mark and CLASS to signify compliance with the appropriate Class Rules.

Correction to: Guidance on the Classification of Offshore Containers

Correction to:
A. A. Olsen, *Guidance on the Classification of Offshore Containers*,
Synthesis Lectures on Ocean Systems Engineering,
https://doi.org/10.1007/978-3-031-74857-8

This book contains overlap in text with the previously published content [1] that was inadvertently omitted. The authors failed to attribute the reference [1]. The authors have now obtained permission to re-use this content from the American Bureau of Shipping

Where [1] is: American Bureau of Shipping (2024), Rules and Guides https://ww2.eagle.org/en/rules-and-resources/rules-and-guides.html

---

The updated version of this book can be found at
https://doi.org/10.1007/978-3-031-74857-8

Annex

Determination of Working Load Limit of the WLL$_{min}$ Lifting Set

Container rating (R) kg	Dynamic factor/enhancement factor	Minimum required working load limit of the lifting set WLL$_{min}$ tonnes
500	–	7.00
1000	–	7.00
1500	–	7.00
2000	3.500	7.00
2500	2.880	7.20
3000	2.600	7.80
3500	2.403	8.41
4000	2.207	8.83
4500	2.067	9.30
5000	1.960	9.80
5500	1.873	10.30
6000	1.766	10.60
6500	1.733	11.26
7000	1.700	11.90
7500	1.666	12.50
8000	1.633	13.07
8500	1.600	13.6
9000	1.567	14.1

(continued)

© The Editor(s) (if applicable) and The Author(s), under exclusive license
to Springer Nature Switzerland AG 2025
A. A. Olsen, *Guidance on the Classification of Offshore Containers*, Synthesis Lectures
on Ocean Systems Engineering, https://doi.org/10.1007/978-3-031-74857-8

(continued)

Container rating (R) kg	Dynamic factor/enhancement factor	Minimum required working load limit of the lifting set WLL$_{min}$ tonnes
9500	1.534	14.57
10,000	1.501	15.01
10,500	1.479	15.53
11,000	1.457	16.02
11,500	1.435	16.5
12,000	1.413	16.95
12,500	1.391	17.38
13,000	1.368	17.79
13,500	1.346	18.18
14,000	1.324	18.54
14,500	1.302	18.88
15,000	1.28	19.20
15,500	1.267	19.64
16,000	1.254	20.06
16,500	1.24	20.47
17,000	1.227	20.86
17,500	1.214	21.24
18,000	1.201	21.61
18,500	1.188	21.97
19,000	1.174	22.31
19,500	1.161	22.64
20,000	1.148	22.96
20,500	1.143	23.44
21,000	1.139	23.92
21,500	1.135	24.39
22,000	1.13	24.86
22,500	1.126	25.33
23,000	1.121	25.79
23,500	1.117	25.25
24,000	1.112	26.7
24,500	1.108	27.15
25,000	1.104	27.59